# Lecture Notes in Artificial Intelligence  11834

Subseries of Lecture Notes in Computer Science

More information about this series at http://www.springer.com/series/1244

Saša Pekeč · Kristen Brent Venable (Eds.)

# Algorithmic
# Decision Theory

6th International Conference, ADT 2019
Durham, NC, USA, October 25–27, 2019
Proceedings

 Springer

*Editors*
Saša Pekeč
Fuqua School of Business
Duke University
Durham, NC, USA

Kristen Brent Venable
Florida Institute for Human and Machine
Cognition (IHMC)
Pensacola, FL, USA

ISSN 0302-9743        ISSN 1611-3349  (electronic)
Lecture Notes in Artificial Intelligence
ISBN 978-3-030-31488-0        ISBN 978-3-030-31489-7  (eBook)
https://doi.org/10.1007/978-3-030-31489-7

LNCS Sublibrary: SL7 – Artificial Intelligence

This Springer imprint is published by the registered company Springer Nature Switzerland AG
The registered company address is: Gewerbestrasse 11, 6330 Cham, Switzerland

# Preface

The 6th International Conference on Algorithmic Decision Theory (ADT 2019), held in October 2019 at Duke University's Fuqua School of Business (Durham, NC, USA), has continued in the tradition established by previous ADT conferences in providing a unique opportunity for scientific exchange among researchers and practitioners coming from Computer Science, Economics, and Operations Research.

ADT 2019 provided a multi-disciplinary forum for sharing knowledge in these areas with a special focus on algorithmic issues in decision theory. Previous ADT conferences were held in Venice, Italy (2009); Piscataway, NJ, USA (2011); Brussels, Belgium (2013); Lexington, KY, USA (2015), and Luxembourg (2017).

Contributions to ADT 2019 addressed key topics related to decision theory, such as, preference elicitation and aggregation, voting, games, decision making over complex domains, social networks, and applications such as decision making for unmanned vehicles, refugee allocation, and kidney exchanges.

The program also included three exceptional invited talks by David Pennock (Microsoft Research), Ariel Procaccia (CMU), and Francesca Rossi (IBM Research).

The papers in this volume were presented at ADT 2019. Each submission received three reviews by Program Committee (PC) members in a double-blind fashion. The PC selected 10 submission as full papers and 7 short papers. Given that 31 submission were received, the acceptance rate was approximately 54%. Two other accepted papers were submitted only for presentation at the conference and are not contained in the proceedings:

- Rupert Freeman, David Pennock, Dominik Peters, and Jennifer Wortman Vaughan: "Truthful Aggregation of Budget Proposals,"
- Aleksandr M. Kazachkov and Shai Vardi: "On Tanking and Competitive Balance."

We thank the authors for submitting and presenting their high quality recent research results. We are also grateful to Lirong Xia for enhancing the program with an exceptional tutorial.

We would like to thank the Program Committee and their additional reviewers for their contribution to the program and to the paper selection process. We also thank Alexis Tsoukiàs for his invaluable advice and support, as well as the ADT 2017 organizers (in particular Jörg Rothe), CNRS (France), and Springer. We also appreciated using the EasyChair platform for organizing the submission and reviewing process.

Finally, our gratitude goes to our sponsors for their generous support: Computer Science Department and Fuqua School of Business at Duke University, and EURO working group on Preference Handling.

July 2019

Saša Pekeč
Kristen Brent Venable

# Organization

## Organizing Committee

| | |
|---|---|
| Vince Conitzer | Duke University, USA |
| Saša Pekeč (Chair) | Duke University, USA |
| Alexis Tsoukiàs | LAMSADE, Université Paris-Dauphine, France |
| Kristen Brent Venable | IHMC, UWF, USA |

## Program Committee

| | |
|---|---|
| Ali Abbas | University of Southern California, USA |
| Alessandro Arlotto | Duke University, USA |
| Haris Aziz | Data61/CSIRO, UNSW, Australia |
| Sylvain Bouveret | LIG - Grenoble INP, Université Grenoble-Alpes, France |
| Robert Bredereck | TU Berlin, Germany |
| Katarina Cechlarova | PF UPJS Kosice, Slovakia |
| Edith Elkind | University of Oxford, UK |
| Piotr Faliszewski | AGH University of Science and Technology, Poland |
| Judy Goldsmith | University of Kentucky, USA |
| Umberto Grandi | University of Toulouse, France |
| Jérôme Lang | CNRS, LAMSADE, Université Paris-Dauphine, France |
| Ali Makhdoumi | Duke University, USA |
| Azarakhsh Malekian | University of Toronto, Canada |
| Nicholas Mattei | Tulane University, USA |
| Saša Pekeč (Co-chair) | Duke University, USA |
| Hans Peters | Maastricht University, The Netherlands |
| Maria Silvia Pini | University of Padova, Italy |
| Marc Pirlot | Université de Mons, Belgium |
| Luca Rigotti | University of Pittsburgh, USA |
| Fred Roberts | Rutgers University, USA |
| Francesca Rossi | IBM Research New York, USA |
| Jörg Rothe | Universität Düsseldorf, Germany |
| Alexis Tsoukiàs | CNRS, LAMSADE, France |
| Kristen Brent Venable (Co-chair) | IHMC, UWF, USA |
| Paolo Viappiani | CNRS, LIP6, Université Pierre et Marie Curie, France |
| Toby Walsh | UNSW, Australia |
| Gerhard J. Woeginger | RWTH Aachen University, Germany |
| Stefan Woltran | Vienna University of Technology, Austria |
| Lirong Xia | RPI, USA |
| Saša Zorc | University of Virginia, USA |

# Contents

# Combining Local Search and Elicitation for Multi-Objective Combinatorial Optimization

Nawal Benabbou, Cassandre Leroy, Thibaut Lust[(✉)], and Patrice Perny

Sorbonne Université, CNRS, LIP6, 75005 Paris, France
{nawal.benabbou,cassandre.leroy,thibaut.lust,patrice.perny}@lip6.fr

**Abstract.** In this paper, we propose a general approach based on local search and incremental preference elicitation for solving multi-objective combinatorial optimization problems with imprecise preferences. We assume that the decision maker's preferences over solutions can be represented by a parameterized scalarizing function but the parameters are initially not known. In our approach, the parameter imprecision is progressively reduced by iteratively asking preference queries to the decision maker (1) before the local search in order to identify a promising starting solution and (2) during the local search but only when preference information are needed to discriminate between the solutions within a neighborhood. This new approach is general in the sense that it can be applied to any multi-objective combinatorial optimization problem provided that the scalarizing function is linear in its parameters (e.g., a weighted sum, an OWA aggregator, a Choquet integral) and that a (near-)optimal solution can be efficiently determined when preferences are precisely known. For the multi-objective traveling salesman problem, we provide numerical results obtained with different query generation strategies to show the practical efficiency of our approach in terms of number of queries, computation time and gap to optimality.

**Keywords:** Multi-objective combinatorial optimization ·
Local search · Preference elicitation · Minimax regret ·
Traveling salesman problem

## 1  Introduction

Designing efficient preference elicitation procedures to support decision making in combinatorial domains is one of the hot topics of algorithmic decision theory. On non-combinatorial domains, various model-based approaches are already available for preference learning. The elicitation process consists in analyzing preference statements provided by the decision maker (DM) to assess the parameters of the decision model and determine an optimal solution (see, e.g., [5,7,9,10,14,31]). Within this stream of work, incremental approaches are of special interest because they aim to analyze the set feasible solutions to identify the

© Springer Nature Switzerland AG 2019
S. Pekeč and K. B. Venable (Eds.): ADT 2019, LNAI 11834, pp. 1–16, 2019.
https://doi.org/10.1007/978-3-030-31489-7_1

critical preference information needed to find the optimal alternative. By a careful selection of preference queries, they make it possible to determine the optimal choice within large sets, using a reasonably small number of questions, see e.g., [10] for an example in a Bayesian setting, and [7] for another approach based on a progressive reduction of the uncertainty attached to the model parameters.

The incremental approach was efficiently used in various decision contexts such as multiattribute utility theory or multicriteria decision making [8,30,34], decision making under risk [10,17,26,32] and collective decision making [23]. However, extending these approaches for decision support on combinatorial domains is more challenging due to the implicit definition of the set of solutions and the huge number of feasible solutions. In order to overcome this problem, several contributions aim to combine standard search procedures used in combinatorial optimization with incremental preference elicitation. Examples can be found in various contexts such as constraint satisfaction [15], committee election [3], matching [13], sequential decision making under risk [28,33], fair multiagent optimization [6] and multicriteria optimization [1,19].

In multicriteria optimization, the search procedure combines the implicit enumeration of Pareto-optimal solutions with preferences queries allowing a progressive reduction of the uncertainty attached to the parameters of the preference aggregation model, in order to progressively focus the exploration on the most attractive solutions. Various attempts to interleave incremental preference elicitation methods and constructive algorithms have been proposed. The basic principle consists in constructing the optimal solution from optimal sub-solutions using the available preference information, and to ask new preference information when necessary. This has been tested for greedy algorithms, for dynamic programming, for $A^*$ and branch-and-bound search, see [4] for a synthesis.

In this paper we explore another way by considering non-constructive algorithms. We propose to interleave the elicitation with local search for multicriteria optimization. To illustrate our purpose, let us consider the following example:

**Example 1.** *Let us consider the instance of the multi-objective traveling salesman problem (TSP) depicted in Fig. 1, including 5 nodes and two additive cost functions to be minimized (one looks for a cycle passing exactly once through each node of the graph and minimizing costs). Let us start a local search from*

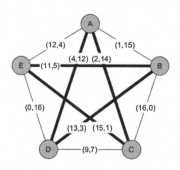

**Fig. 1.** An instance of the TSP with two criteria.

*the boldface tour $x_0 = ADBECA$ whose cost vector is $(45, 35)$, using a neighborhood definition based on the simple exchange of two consecutive nodes. Among the neighbors of $x_0$, there is $x'_0 = ABDECA$ whose cost vector is $(31, 49)$. Assume that the DM declares that $x_0$ is better than $x'_0$ (denoted $x_0 \succ x'_0$).*

Let us show what could be a local search on this instance using partial preference information interpreted under a linear model assumption:

**Using a Linear Model.** We assume here that DM's preferences can be represented by a linear model of the form: $f_\omega(y_1, y_2) = \omega y_1 + (1 - \omega)y_2$ for some unknown $\omega \in (0, 1)$, where $(y_1, y_2)$ is the cost vector associated to a feasible tour. In this case the $x_0 \succ x'_0$ condition implies $f_\omega(45, 35) < f_\omega(31, 49)$ and therefore $\omega \in (0, 1/2)$. After this restriction of the set of possible values for $\omega$ it can easily be checked that the optimal neighbor of $x_0$ is solution $x_1 = ADBCEA$ of cost $(60, 20)$. Then by exploring the neighborhood of $x_1$ it can easily be checked that no other solution can improve $x_1$ given that $\omega < 1/2$. We get a local optimum which is actually the optimal solution for this instance.

We can see here that, under the linear model assumption, an optimal solution has been obtained using a single preference query. However, the linear model is not always suitable. For example, when one looks for a compromise solution between the two criteria, one could prefer resorting to a decision model favoring the generation of solutions having a balanced cost vector. For this reason we consider now another elicitation session using a non-linear weighted aggregation function commonly used to control the balance between criterion values, namely the *Ordered Weighted Average* (OWA, [22, 35]).

**Using the OWA Model.** Now, let us assume that the DM's preferences are represented by a non-linear model of the form: $f_\omega(y_1, y_2) = \omega \max\{y_1, y_2\} + (1 - \omega) \min\{y_1, y_2\}$ for some unknown $\omega \in (0, 1)$. In this case the $x_0 \succ x'_0$ condition implies $45\omega + 35(1 - \omega) < 49\omega + 31(1 - \omega)$ and therefore $\omega \in (1/2, 1)$ (note that although OWA is not linear in $y$, it is linear in $\omega$ and therefore any preference statement translates into a linear constraint on $\omega$). After this restriction of the set of possible values for $\omega$ it can easily be checked that the optimal neighbor of $x_0$ is solution $x_2 = ABECDA$ whose cost vector is $(40, 40)$. Then, by exploring the neighborhood of $x_2$, it can easily be checked that no other solution can improve $x_2$ given that $\omega > 1/2$. We obtain a local optimum which is actually the OWA-optimal solution for this instance, given the restriction on $\omega$.

These simple executions of local search using partial preference information show the potential of interactive local search combining local exploration of neighborhoods and model-based preference elicitation. For a given class of preference models, the successive answers from the DM to preference queries make it possible to progressively reduce the set of possible parameters and to discriminate the solutions belonging to the neighborhood of solutions found so far.

The combination of preference elicitation has several specific advantages. In particular, preference elicitation is based on very simple queries because they involve neighbor solutions that are cognitively simpler to compare than pairs of solutions varying in all aspects. Moreover, preference queries only involve complete solutions. This provides two advantages: (1) solutions are easier to compare, and (2) no independence assumption is required in the definition of preferences (no need to reason on partial descriptions under the assumption that preferences hold everything all being equal). The latter point is of special interest when preferences are represented by *non-linear* decision models. With such models, the cost of partial solutions is a poor predictor of the actual cost of their extensions. This seriously reduces possibilities of pruning sub-solutions in constructive algorithms. Since we consider only complete solutions in local search, this problem vanishes. Another interest of studying incremental elicitation approaches in local search is to tackle preference-based combinatorial optimization problems for which no efficient exact algorithm is known.

The paper is organized as follows. Section 2 introduces some preliminary background and notations. Then, we present a general interactive local search in Sect. 3. In Sect. 4 we further specify our approach for application to the multi-objective TSP and provide numerical tests showing the practical efficiency of the proposed incremental elicitation process.

## 2    Background and Notations

In this section, we present the necessary background on multi-objective combinatorial optimization and regret-based incremental elicitation.

### 2.1    Multi-Objective Combinatorial Optimization

We consider a multi-objective combinatorial optimization (MOCO) problem with $n$ objectives/criteria $y_i, i \in \{1, \ldots, n\}$, that need to be minimized. This problem can be formulated as follows: $\text{minimize}_{x \in \mathcal{X}} (y_1(x), \ldots, y_n(x))$ where $\mathcal{X}$ is the feasible set in the decision space (e.g., for the TSP, $\mathcal{X}$ is the set of all Hamiltonian cycles). In this problem, any solution $x \in \mathcal{X}$ is associated with a vector $y(x) = (y_1(x), \ldots, y_n(x)) \in \mathbb{R}^n$ that gives its evaluations on all criteria. Solutions are usually compared through their images in the objective space (also called points) using the *Pareto dominance* relation: point $a = (a_1, \ldots, a_n) \in \mathbb{R}^n$ is said to *Pareto dominate* point $b = (b_1, \ldots, b_n) \in \mathbb{R}^n$ (denoted by $a \prec_P b$) if and only if $a_i \leq b_i$ for all $i \in \{1, \ldots, n\}$ and $a_i < b_i$ for some $i \in \{1, \ldots, n\}$. A solution $x \in \mathcal{X}$ is said to be *efficient* if there is no other feasible solution $x' \in \mathcal{X}$ such that $y(x') \prec_P y(x)$ and the set $\mathcal{X}_E$ of all efficient solutions is called the *efficient set* (their images are respectively called *non-dominated point* and *Pareto front*).

We assume here that the DM needs to select a single solution. Without any preference information, we only know that her preferred solution is an element of the efficient set. However, it is well-known that the number of efficient solutions (and the number of non-dominated points) can be exponential in the size of the problem (e.g., [18] for the multicriteria spanning tree problem); in such situations, identifying the Pareto front is not enough to help the DM in making a decision. One way to address this issue is to reduce the size of the Pareto front by constructing a "well-represented" set; for instance, this set can be obtained by dividing the objective space into different regions (e.g., [20]) or by using some approximate dominance relation (e.g., $\epsilon$-dominance [21]). However, in situations where the DM needs to select only one solution, it seems more appropriate to refine the Pareto dominance with preferences in order to determine the optimal solution according to the DM's subjective preferences.

In this work, we assume that the DM's subjective preferences can be represented by a parameterized scalarizing function $f_\omega$ that is linear in its preference parameters $\omega$. For example, function $f_\omega$ can be a weighted sum (i.e., $f_\omega(a) = \sum_{i=1}^{n} \omega_i a_i$), an OWA aggregator ($f_\omega(a) = \sum_{i=1}^{n} \omega_i a_{(i)}$ where $a_{(i)} \geq \ldots \geq a_{(n)}$ are the components of $a$ sorted in non-increasing order, see e.g. [35]) or even a Choquet integral with capacity $\omega$ (see e.g. [11, 16]). In this context, solution $x \in \mathcal{X}$ is preferred to solution $x' \in \mathcal{X}$ by the DM if and only if $f_\omega(y(x)) \leq f_\omega(y(x'))$. Thus any solution $x \in \mathcal{X}$ that minimizes function $f_\omega$ is optimal according to the DM's preferences.

## 2.2   Regret-Based Incremental Elicitation

For the purpose of elicitation, we assume that preference parameters $\omega$ are not known initially. More precisely, we are given a (possibly empty) set $\Theta$ of preference statements of type $(a, b) \in \mathbb{R}^n \times \mathbb{R}^n$, meaning that the DM prefers point $a$ to point $b$, and we consider the set $\Omega_\Theta$ of all parameters $\omega$ that are compatible with $\Theta$ the available preference information. Formally, set $\Omega_\Theta$ is defined by $\Omega_\Theta = \{\omega : \forall (a, b) \in \Theta, f_\omega(a) \leq f_\omega(b)\}$. Note that $\Omega_\Theta$ is a convex polyhedron since function $f_\omega$ is assumed to be linear in its parameters $\omega$.

Our goal now is to determine the most promising solution under the parameter imprecision characterized by $\Omega_\Theta$. To this aim, we consider the minimax regret approach (e.g., [7]) which is based on the following definitions:

**Definition 1 (Pairwise Max Regret).** *The Pairwise Max Regret (PMR) of solution $x \in \mathcal{X}$ with respect to solution $x' \in \mathcal{X}$ is:*

$$PMR(x, x', \Omega_\Theta) = \max_{\omega \in \Omega_\Theta} \left\{ f_\omega(y(x)) - f_\omega(y(x')) \right\}$$

By definition, $PMR(x, x', \Omega_\Theta)$ is the worst-case loss when recommending solution $x$ instead of solution $x'$ to the DM[1].

**Definition 2 (Max Regret).** *The Max Regret (MR) of solution $x \in \mathcal{X}$ is:*

$$MR(x, \mathcal{X}, \Omega_\Theta) = \max_{x' \in \mathcal{X}} PMR(x, x', \Omega_\Theta)$$

In other words, $MR(x, \mathcal{X}, \Omega_\Theta)$ is the worst-case loss when choosing $x$ instead of any other solution $x' \in \mathcal{X}$. Finally, the minimax reget is defined as follows:

**Definition 3 (Minimax Regret).** *The MiniMax Regret (MMR) is:*

$$MMR(\mathcal{X}, \Omega_\Theta) = \min_{x \in \mathcal{X}} MR(x, \mathcal{X}, \Omega_\Theta)$$

---

[1] Note that $PMR(x, x', \Omega_\Theta)$ values can be computed using a LP solver since $\Omega_\Theta$ is described by linear constraints and $f_\omega$ is linear in its parameters $\omega$.

A solution $x^* \in \mathcal{X}$ is optimal according to the minimax regret decision crite-
rion if $x^*$ achieves the minimax regret, i.e. if $x^* \in \arg\min_{x \in \mathcal{X}} MR(x, \mathcal{X}, \Omega_\Theta)$.
Recommending such a solution guarantees that the worst-case loss is mini-
mized (given the imprecision surrounding the DM's preferences). Moreover, if
$MMR(\mathcal{X}, \Omega_\Theta) = 0$, then we know that any optimal solution for the minimax
regret criterion is necessarily optimal according to the DM's preferences.

Note that we have $MMR(\mathcal{X}, \Omega_{\Theta'}) \leq MMR(\mathcal{X}, \Omega_\Theta)$ for any set $\Theta' \supseteq \Theta$, as
already observed in previous works (see e.g., [5]). Thus the general principle of
regret-based incremental elicitation is to iteratively collect preference information by asking preference queries to the DM until $MMR(\mathcal{X}, \Omega_\Theta)$ drops below
a given tolerance threshold $\delta \geq 0$ (representing the maximum allowable gap to
optimality); if we set $\delta = 0$, then we obtain the (optimal) preferred solution at
the end of the execution.

## 3   An Interactive Local Search Algorithm

For MOCO problems, regret-based incremental elicitation may induce prohibitive computation times since it may require to compute the pairwise max
regrets for all pairs of distinct solutions in $\mathcal{X}$ (see Definitions 2 and 3). This
observation has led a group of researchers to propose a new approach consisting
in combining regret-based incremental elicitation and search by asking preference queries during the construction of the (near-)optimal solution (e.g. [2]).
In this paper, we combine incremental elicitation and search in a different way.
More precisely, we propose an interactive local search procedure that generates
preference queries (1) before the local search to determine a promising starting
point and (2) during the local search to help identifying the best solution within
a neighborhood.

Our interactive algorithm takes as input a MOCO problem $P$, two thresholds
$\delta = (\delta_1, \delta_2), (\delta_1, \delta_2 \geq 0)$, a scalarizing function $f_\omega$ with unknown parameters $\omega$,
an initial set of preference statements $\Theta$ (possibly empty), and $m$ the number of
possible starting solutions (generated at the beginning of the procedure). First,
our algorithm identifies a promising starting solution as follows:

1. A set of $m$ admissible preference parameters $\omega^k$, $k \in \{1, \ldots, m\}$, are randomly
   generated within $\Omega_\Theta$.
2. Then, for every $k \in \{1, \ldots, m\}$, a (near-)optimal solution is determined for
   the *precise* scalarizing function $f_{\omega^k}$ using an existing efficient algorithm. Let
   $X_0$ be the set of generated solutions.
3. Finally, preference queries are generated in order to discriminate between the
   solutions in $X_0$. More precisely, while $MMR(X_0, \Omega_\Theta) > \delta_1$, the DM is asked
   to compare two solutions $x, x' \in X_0$ and the set of admissible parameters
   is updated by inserting the constraint $f_\omega(x) \leq f_\omega(x')$ (or $f_\omega(x) \geq f_\omega(x')$
   depending on her answer); once $MMR(X_0, \Omega_\Theta)$ drops below $\delta_1$, the starting
   solution $x^*$ is arbitrarily chosen in $\arg\min_{x \in X_0} MR(x, X_0, \Omega_\Theta)$.

Then, our algorithm moves from solution to solution by considering local
improvements. More precisely, it iterates as follows:

1. Firstly, a set $X^*$ of solutions is generated from $x^*$ using a neighborhood function defined on the search space; we add $x^*$ to $X^*$ and remove from $X^*$ any solution that is Pareto-dominated by another solution in this set.
2. Secondly, while $MMR(X^*, \Omega_\Theta) > \delta_2$, the DM is asked to compare two solutions $x, x' \in X^*$ and $\Omega_\Theta$ is restricted by inserting the constraint $f_\omega(x) \leq f_\omega(x')$ (or $f_\omega(x) \geq f_\omega(x')$).
3. Finally, if $MR(x^*, X^*, \Omega_\Theta) > \delta_2$ holds, solution $x^*$ is then replaced by a neighbor solution minimizing the max regret in $X^*$; otherwise, the algorithm stops by returning solution $x^*$.

Our algorithm is called ILS for Interactive Local Search and is summarized in Algorithm 1.

---

**Algorithm 1. ILS**

---

**IN** ↓ $P$: a MOCO problem; $\delta_1$, $\delta_2$: thresholds; $f_\omega$: an aggregator with unknown parameters; $\Theta$: a set of preference statements; $m$: number of initial solutions.

--| Initialization of the admissible parameters:
$\Omega_\Theta \leftarrow \{\omega : \forall(a, b) \in \Theta, f_\omega(a) \leq f_\omega(b)\}$
--| Generation of $m$ initial solutions:
$X_0 \leftarrow \texttt{Select\&Optimize}(P, \Omega_\Theta, m)$
--| Determination of the starting solution:
**while** $MMR(X_0, \Omega_\Theta) > \delta_1$ **do**
   --| Ask the DM to compare two solutions in $X_0$:
   $(x, x') \leftarrow \texttt{Query}(X_0)$
   --| Update preference information:
   $\Theta \leftarrow \Theta \cup \{(y(x), y(x'))\}$
   $\Omega_\Theta \leftarrow \{\omega : \forall(a, b) \in \Theta, f_\omega(a) \leq f_\omega(b)\}$
**end while**
$x^* \leftarrow \texttt{Select}(\arg\min_{x \in X_0} MR(x, X_0, \Omega_\Theta))$
--| Interactive Local Search:
improve $\leftarrow$ **true**
**while** improve **do**
   $X^* \leftarrow \texttt{Neighbors}(P, x^*) \cup \{x^*\}$
   **while** $MMR(X^*, \Omega_\Theta) > \delta_2$ **do**
      --| Ask the DM to compare two solutions in $X^*$:
      $(x, x') \leftarrow \texttt{Query}(X^*)$
      --| Update preference information:
      $\Theta \leftarrow \Theta \cup \{(y(x), y(x'))\}$
      $\Omega_\Theta \leftarrow \{\omega : \forall(a, b) \in \Theta, f_\omega(a) \leq f_\omega(b)\}$
   **end while**
   --| Move to another solution:
   **if** $MR(x^*, X^*, \Omega_\Theta) > \delta_2$ **then**
      $x^* \leftarrow \texttt{Select}(\arg\min_{x \in X^*} MR(x, X^*, \Omega_\Theta))$
   **else**
      improve $\leftarrow$ **false**
   **end if**
**end while**
**return** $x^*$

---

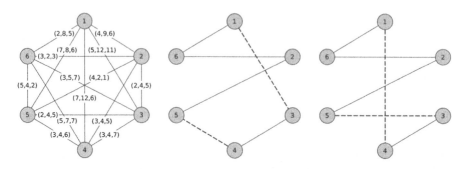

**Fig. 2.** The graph on the left side of the figure represents an instance of the 3-objective TSP with 6 vertices. The two others graphs give an example of a 2-opt movement: the dashed edges of the cycle in the middle are deleted and then replaced by the dashed edges in the right side of the figure.

Note that procedures `Select&Optimize` and `Neighbors` depend on the problem considered; for instance, for the multicriteria spanning tree problem with a weighted sum model, the optimization part can be performed using Prim algorithm [27] and the neighborhood function can be defined by edge swaps. Note also that procedure `Query`$(X)$ can implement any query generation strategy that selects two solutions in $X$ and asks the DM to compare them; in the numerical section, we propose and compare different query generation strategies.

It is well-known that local search is a heuristic search that may stuck at a locally optimal point that is not globally optimal; the problem obviously remains when using our interactive local search. However, it is worth noting that our algorithm with $\delta_2 = 0$ provides the same performance guarantees as the corresponding local search algorithm with precise preferences. To give an example, when using the 2-opt neighborhood function [12], our algorithm approximately solves the TSP within a differential-approximation ratio bounded above by $1/2$ (see [24] for further details); in the numerical section, we will see that the error is even lower in practice.

For illustration purposes, we now present an execution of our algorithm on a small instance of the multi-objective TSP:

**Example 2.** *Consider the 3-objective TSP with 6 vertices defined by Fig. 2. In this problem, the set $\mathcal{X}$ of feasible solutions is the set of all Hamiltonian cycles, i.e. cycles that include every node exactly once. We now apply ILS algorithm with $\delta = (0,0)$ on this instance considering the neighborhood function defined by 2-opt swaps [12]; in other words, the neighbors of cycles are all the cycles that can be obtained by deleting two edges and adding two other edges from the graph (see Fig. 2 for an example).*

*We assume here that the DM's preferences can be represented by a weighted sum with the hidden weight $\omega^* = (0.2, 0.1, 0.7)$ and we start the execution with an empty set of preference statements (i.e. $\Theta = \emptyset$). Hence $\Omega_\Theta$ is initially the set of all weighting vectors $\omega = (\omega_1, \omega_2, \omega_3) \in [0,1]^3$ such that $\omega_1 + \omega_2 + \omega_3 = 1$. In Fig. 3, we represent $\Omega_\Theta$ by the triangle $ABC$ in the space $(\omega_1, \omega_2)$, $\omega_3$ being implicitly defined by $\omega_3 = 1 - \omega_1 - \omega_2$.*

**Identification of a Promising Starting Solution:** *First, we generate $m = 2$ weighting vectors $\omega^1$ and $\omega^2$ at random and we determine then the corresponding optimal solutions $x^1$ and $x^2$. If $\omega^1 = (0.6, 0.3, 0.1)$ and $\omega^2 = (0.3, 0.6, 0.1)$, we obtain the following evaluation: $y(x^1) = (19, 34, 30)$ and $y(x^2) = (21, 32, 27)$. Let $X_0 = \{x^1, x^2\}$. Since $MMR(X_0, \Omega_\Theta) = 2 > \delta_1$, we ask the DM to compare $x^1$ and $x^2$. Since $f_{\omega^*}(y(x^1)) = 28.2 > f_{\omega^*}(y(x^2)) = 26.3$, the DM prefers solution $x^2$ to $x^1$. Therefore we set $\Theta = \{((21, 32, 27), (19, 34, 30))\}$ and $\Omega_\Theta$ is restricted by imposing the constraint $f_\omega(y(x^2)) \leq f_\omega(y(x^1))$, i.e. $\omega_2 \leq -5\omega_1 + 3$ (see Fig. 4 where $\Omega_\Theta$ is represented by $ABDE$). Now we have $MMR(X_0, \Omega_\Theta) = MR(x^2, X_0, \Omega_\Theta) = 0 \leq \delta_1$, and therefore $x^2$ is chosen to be the starting solution (i.e. $x^* = x^2$).*

**Local Search:** *At the first iteration step, three neighbors of $x^*$ are Pareto non-dominated, and the set $X^*$ contains four solutions, denoted by $x^1$, $x^2(= x^*)$, $x^3$ and $x^4$ evaluated as follows: $y(x^1) = (23, 34, 26)$, $y(x^2) = (21, 32, 27)$, $y(x^3) = (19, 34, 30)$ and $y(x^4) = (20, 31, 30)$. Since $MMR(X^*, \Omega_\Theta) = 1 > \delta_2$, we ask the DM to compare two solutions in $X^*$, say $x^1$ and $x^*$. As $f_{\omega^*}(y(x^1)) = 26.2 < f_{\omega^*}(y(x^*)) = 26.3$, the DM prefers $x^1$ to $x^*$. Therefore we obtain $\Theta = \{((21, 32, 27), (19, 34, 30)), ((23, 34, 26), (21, 32, 27))\}$ and $\Omega_\Theta$ is restricted by the linear constraint $f_\omega(y(x^1)) \leq f_\omega(y(x^*))$, i.e. $\omega_2 \leq -\omega_1 + 1/3$ (see Fig. 5 where $\Omega_\Theta$ is represented by $AGF$). Then we stop asking queries at this step since we have $MMR(X^*, \Omega_\Theta) = MR(x^1, X^*, \Omega_\Theta) = 0 \leq \delta_2$. We move from $x^* = x^2$ to solution $x^1$ for the next step (i.e., we now set $x^* = x^1$).*

*At the second iteration step, $X^*$ only includes three solutions denoted by $x^1(= x^*)$, $x^2$ and $x^3$ with $y(x^1) = (23, 34, 26)$, $y(x^2) = (21, 32, 27)$ and $y(x^3) = (19, 33, 31)$. Since $MMR(X^*, \Omega_\Theta) = 0 \leq \delta_2$, no query is generated at this step. Moreover, $MR(x^*, X^*, \Omega_\Theta) = 0 \leq \delta_2$ (that is $x^* \in \arg\min_{x \in X^*} MR(x, X^*, \Omega_\Theta)$) and $x^*$ is thus a local optimum (variable improve is set to **false** and the while loop ends). Therefore, after two iteration steps, ILS algorithm stops by returning the solution $x^* = x^1$ which is the preferred solution in this problem. Note that only two preference queries were needed to discriminate between the 60 feasible solutions (among which 10 are Pareto-optimal).*

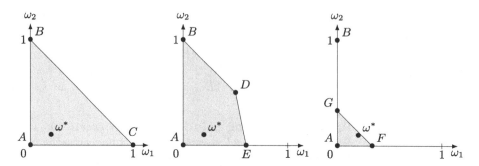

**Fig. 3.** Initial set $\Omega_\Theta$.    **Fig. 4.** $\Omega_\Theta$ after 1 query.    **Fig. 5.** $\Omega_\Theta$ after 2 queries.

To illustrate ILS and the impact of $m$ (the number of initial solutions), we show the evolution of the local search on a randomly generated instance of the bicriteria TSP with 100 cities (see Fig. 6). The left part of the figure ($m = 1$) shows that the neighborhood function enables to go straight to the optimal solution instead of following the Pareto front. However the number of iterations can still be very large when the starting solution is far from the optimal solution in the objective space. The right part of the figure ($m = 2$) shows that the number of iterations can be much reduced when increasing the number of possible starting solutions and selecting the most preferred one.

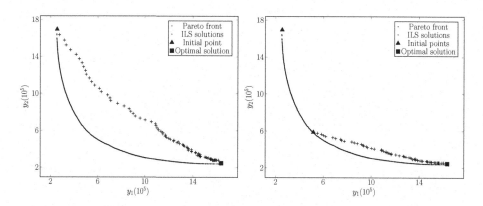

**Fig. 6.** Results obtained for an instance of the 2-criteria TSP.

## 4   Numerical Tests

In this section, we provide numerical results aiming to estimate the performance of our algorithm. In these experiments, we use existing Euclidean instances[2]

---
[2] https://eden.dei.uc.pt/~paquete/tsp/.

of the multi-objective TSP with 25 to 100 cities, and $n = 3$ to 6 objectives. Numerical tests were performed on a Intel Core i7-8550U CPU with 16 GB of RAM, with a program written in C++[3].

### 4.1 Preferences Represented by a Weighted Sum

We assume here that the preferences are represented by a weighted sum $f_\omega$ with imprecise weights, with an empty set of preference statements (i.e. $\Theta = \emptyset$). The answers to queries are simulated using a weighting vector $\omega$ randomly generated before running the algorithm, using the procedure presented in [29], to guarantee a uniform distribution of the weights.

In ILS algorithm, procedure Query($X$) selects two solutions from $X$ and then asks the DM to compare them. We first want to estimate the impact of the query generation strategy on the performances of ILS algorithm. To do so, we consider the following two query generation strategies:

- **Random:** the set of possibly optimal solutions in $X$ are computed and then two of them are selected at random at each iteration step; note that the set of possibly optimal solutions in $X$ can be determined in polynomial time in the size of $X$ (see e.g., [1]).
- **CSS:** The Current Solution Strategy (CSS) selects a solution $x^* \in X$ that minimizes the max regret $MR(x^*, X, \Omega_\Theta)$ and then the DM is asked to compare solution $x^*$ with one of its adversary's choice (i.e. a solution in $\arg\max_{x \in X} PMR(x^*, x, \Omega_\Theta)$) [7].

These strategies are compared in terms of computation time (given in seconds), number of generated queries and the error (expressed in terms of percentage from the optimal solution). Results averaged over 100 runs are given in Table 1 for the instance with 50 cities. In this table, "/" means that the timeout is exceeded (the timeout is set to 900 seconds).

**Table 1.** Comparison between CSS and Random strategies, 50 cities, $\delta = (0,0)$.

| $n$ | $m$ | CSS time | CSS queries | CSS error | Random time | Random queries | Random error |
|---|---|---|---|---|---|---|---|
| 4 | 10 | 28.52 | 35.90 | 1.96 | 15.33 | 77.54 | 1.90 |
| 4 | 50 | 8.65 | 27.31 | 1.11 | 317.11 | 380.18 | 0.92 |
| 4 | 100 | 6.71 | 24.34 | 0.69 | 894.93 | 375.83 | 0.70 |
| 6 | 10 | 537.07 | 85.67 | 2.06 | 67.91 | 168.71 | 2.31 |
| 6 | 50 | 133.08 | 68.45 | 1.64 | 899.39 | 338.98 | 1.49 |
| 6 | 100 | 43.06 | 54.40 | 1.35 | > 900 | / | / |

---

[3] PMRs values are computed using CPLEX Optimizer (https://www.ibm.com/analytics/cplex-optimizer) and the optimization part of Select&Optimize is performed by the exact TSP solver Concorde (http://www.math.uwaterloo.ca/tsp/concorde).

First, we see that the query strategy has an important impact on the quality of the results: with the random strategy the number of queries and computation time are much higher than with the CSS strategy, showing that preference queries must be carefully chosen when designing incremental elicitation methods. Then we see that ILS with CSS achieves better results when increasing the number of possible initial solutions, in terms of computation times, queries and error (as the selected starting solution is becoming closer to the best solution). Although the error is very low (about less than 2%) for both strategies, the number of queries is quite high for instances with 6 criteria (at least 54 queries). This is due to the fact that we use the tolerance thresholds $\delta = (0,0)$.

We now compare the results obtained with $\delta = (0,0)$ and $\delta = (0.1, 0.4)$ (see the left part of Table 2); we set $\delta_1 < \delta_2$ since the starting point selection has a significant impact on local search performances. We also give the results obtained by ILS when a ranking of the objectives can be provided by the DM prior to the search (see the right part of Table 2). We vary $n$ the number of criteria between 3 and 6, and we set $m$ the number of initial solutions to 100.

In Table 2, we see that using strictly positive thresholds enables to reduce both the number of queries and the computation time without having much impact on the error. For instance, for $n = 6$, the error is equal to 1.35% with $\delta = (0,0)$ whereas it is equal to 1.77% with $\delta = (0.1, 0.4)$, while the number of queries is reduced from 54.40 to 32.24. We also remark that knowing the ranking of the objectives allows to further reduce the number of queries (only 23.75 queries for $n = 6$, with an error of 1.51%).

**Table 2.** Performances of ILS combined with CSS, with (left) and without the ranking of the objectives (right), 50 cities, $m = 100$, 100 runs.

| $n$ | $\delta = (0,0)$ | | | $\delta = (0.1, 0.4)$ | | | $\delta = (0,0)$ | | | $\delta = (0.1, 0.4)$ | | |
|---|---|---|---|---|---|---|---|---|---|---|---|---|
| | time | queries | error | time | queries | error | time | queries | error | time | queries | error |
| 3 | 2.67 | 13.50 | 0.20 | 3.81 | 13.65 | 2.01 | 2.10 | 10.53 | 0.20 | 2.30 | 9.66 | 0.80 |
| 4 | 6.71 | 24.34 | 0.69 | 8.32 | 19.68 | 1.84 | 5.01 | 17.01 | 0.66 | 4.89 | 13.39 | 1.13 |
| 5 | 19.96 | 38.38 | 0.96 | 19.30 | 26.21 | 2.08 | 14.81 | 25.08 | 0.95 | 10.97 | 19.24 | 1.38 |
| 6 | 43.06 | 54.40 | 1.35 | 36.25 | 32.24 | 1.77 | 31.19 | 34.52 | 1.36 | 29.68 | 23.75 | 1.51 |

In Fig. 7, we show the evolution of the number of queries according to the number of criteria (left part, 50 cities) and number of cities (right), with $\delta = (0.1, 0.4)$ and $m = 100$. We note that the number of queries evolves more or less linearly according to the number of criteria/cities.

## 4.2   Preferences Represented by an OWA Aggregator

Now we assume that the DM's preferences can be represented by an OWA aggregator, with decreasing weights, favoring well-balanced solutions [35] (as the larger weights are associated with the worst values). Contrary to the weighted sum, when the weights $\omega$ are known, we cannot reduce the multi-objective TSP to

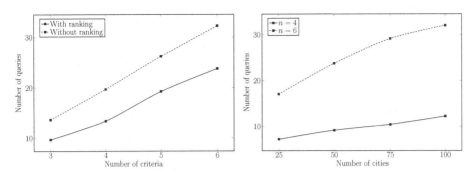

**Fig. 7.** Evolution of the queries according to number of criteria and number of cities.

a single-objective TSP as the OWA aggregator is not a linear operator. Therefore, to obtain the optimal solution with known weights (to be able to compute the error), we have used a well-known linearization of the OWA operator with decreasing weights (see [25]); for information purposes, we also provide the computation time needed when solving the corresponding linear program with the LP-solver (see "LP time").

In Table 3, we give the results obtained by ILS combined with the CSS for the instances with 50 cities, 3 to 6 criteria and $m = 10$ starting solutions (obtained by optimizing 10 randomly generated weighted sum). The results show that the number of queries is much reduced compared to the weighted sum, with an error also around 2%. Moreover, we observe that using our algorithm to solve the problem with unknown weights is much faster than using the LP-solver to obtain the optimal solution when the weights are known. This shows that our algorithm can be used to efficiently solve multi-objective optimization problems with complex decision models, even for problems such that there is no efficient algorithm for the determination of the optimal solution with known preference parameters.

**Table 3.** ILS combined with CSS with OWA, 50 cities, $m = 10$.

| $n$ | LP time | $\delta = (0,0)$ | | | $\delta = (0.1, 0.4)$ | | |
|---|---|---|---|---|---|---|---|
| | | time | queries | error | time | queries | error |
| 3 | 164.80 | 1.14 | 9.96 | 2.31 | 1.19 | 6.16 | 2.41 |
| 4 | 330.17 | 1.28 | 10.57 | 1.98 | 1.06 | 6.13 | 2.00 |
| 5 | 730.19 | 0.98 | 6.40 | 1.17 | 0.83 | 4.09 | 1.42 |
| 6 | 7870.03 | 1.12 | 16.00 | 1.41 | 0.89 | 9.13 | 1.44 |

Finally, in Fig. 8, we illustrate the iterations of ILS when preferences are represented by an OWA operator with decreasing weights (which favors well-balanced solutions).

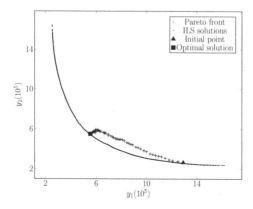

**Fig. 8.** Results obtained for an instance of the 2-criteria TSP with OWA operator.

## 5  Conclusion

In this paper, we have proposed a general approach based on local search and incremental elicitation for solving multi-objective combinatorial optimization problems with unkown preference parameters. We have applied the method to a NP-hard combinatorial optimization problem and we have shown that, by combining the generation of promising starting solutions with an adaptive preference-based local search, we are able to rapidly obtain high quality solutions, even with a non-linear aggregation function like OWA. The approach can be applied to any multi-objective combinatorial optimization problem provided that the scalarizing function used to compare solutions is linear in its parameters.

## References

1. Benabbou, N., Perny, P.: Incremental weight elicitation for multiobjective state space search. In: Proceedings of AAAI 2015, pp. 1093–1098 (2015)
2. Benabbou, N., Perny, P.: On possibly optimal tradeoffs in multicriteria spanning tree problems. In: Proceedings of ADT 2015, pp. 322–337 (2015)
3. Benabbou, N., Perny, P.: Solving multi-agent knapsack problems using incremental approval voting. In: Proceedings of ECAI 2016, pp. 1318–1326 (2016)
4. Benabbou, N., Perny, P.: Interactive resolution of multiobjective combinatorial optimization problems by incremental elicitation of criteria weights. EURO J. Decis. Processes **6**(3), 283–319 (2018)
5. Benabbou, N., Perny, P., Viappiani, P.: Incremental elicitation of Choquet capacities for multicriteria choice, ranking and sorting problems. Artif. Intell. **246**, 152–180 (2017)
6. Bourdache, N., Perny, P.: Active preference elicitation based on generalized Gini functions: application to the multiagent knapsack problem. In: Proceedings of AAAI 2019 (2019)
7. Boutilier, C., Patrascu, R., Poupart, P., Schuurmans, D.: Constraint-based optimization and utility elicitation using the minimax decision criterion. Artif. Intell. **170**(8–9), 686–713 (2006)

8. Braziunas, D., Boutilier, C.: Minimax regret based elicitation of generalized additive utilities. In: Proceedings of UAI 2007, pp. 25–32 (2007)
9. Eric, B., Freitas, N.D., Ghosh, A.: Active preference learning with discrete choice data. In: Advances in Neural Information Processing Systems, pp. 409–416 (2008)
10. Chajewska, U., Koller, D., Parr, R.: Making rational decisions using adaptive utility elicitation. In: Proceedings of AAAI 2000, pp. 363–369 (2000)
11. Choquet, G.: Theory of capacities. Annales de l'Institut Fourier **5**, 31–295 (1953)
12. Croes, G.A.: A method for solving traveling-salesman problems. Oper. Res. **6**(6), 791–812 (1958)
13. Drummond, J., Boutilier, C.: Preference elicitation and interview minimization in stable matchings. In: Proceedings of AAAI 2014, pp. 645–653 (2014)
14. Fürnkranz, J., Hüllermeier, E.: Preference Learning. Springer, Heidelberg (2010). https://doi.org/10.1007/978-3-642-14125-6
15. Gelain, M., Pini, M.S., Rossi, F., Venable, K.B., Walsh, T.: Elicitation strategies for soft constraint problems with missing preferences: properties, algorithms and experimental studies. Artif. Intell. **174**(3), 270–294 (2010)
16. Grabisch, M., Labreuche, C.: A decade of application of the Choquet and Sugeno integrals in multi-criteria decision aid. Ann. Oper. Res. **175**(1), 247–286 (2010)
17. Ha, V., Haddawy, P.: Problem-focused incremental elicitation of multi-attribute utility models. In: Proceedings of UAI 1997, pp. 215–222. Morgan Kaufmann Publishers Inc. (1997)
18. Hamacher, H.W., Ruhe, G.: On spanning tree problems with multiple objectives. Ann. Oper. Res. **52**, 209–230 (1994)
19. Kaddani, S., Vanderpooten, D., Vanpeperstraete, J.-M., Aissi, H.: Weighted sum model with partial preference information: application to multi-objective optimization. Eur. J. Oper. Res. **260**, 665–679 (2017)
20. Karasakal, E., Köksalan, M.: Generating a representative subset of the nondominated frontier in multiple criteria decision making. Oper. Res. **57**(1), 187–199 (2009)
21. Laumanns, M., Thiele, L., Deb, K., Zitzler, E.: Combining convergence and diversity in evolutionary multiobjective optimization. Evol. Comput. **10**(3), 263–282 (2002)
22. Lesca, J., Perny, P.: LP solvable models for multiagent fair allocation problems. ECAI **2010**, 393–398 (2010)
23. Lu, T., Boutilier, C.: Robust approximation and incremental elicitation in voting protocols. IJCAI **2011**, 287–293 (2011)
24. Monnot, J., Paschos, V.T., Toulouse, S.: Approximation algorithms for the traveling salesman problem. Math. Methods Oper. Res. **56**(3), 387–405 (2003)
25. Ogryczak, W., Śliwiński, T.: On solving linear programs with the ordered weighted averaging objective. Eur. J. Oper. Res. **148**(1), 80–91 (2003)
26. Perny, P., Viappiani, P., Boukhatem, A.: Incremental preference elicitation for decision making under risk with the rank-dependent utility model. In: Proceedings of UAI, pp. 597–606 (2016)
27. Prim, R.C.: Shortest connection networks and some generalizations. Bell Syst. Tech. J. **36**, 1389–1401 (1957)
28. Regan, K., Boutilier, C.: Eliciting additive reward functions for Markov decision processes. In: Proceedings of IJCAI 2011, pp. 2159–2164 (2011)
29. Rubinstein, R.Y.: Generating random vectors uniformly distributed inside and on the surface of different regions. Eur. J. Oper. Res. **10**(2), 205–209 (1982)

30. Salo, A., Hämäläinen, R.P.: Preference ratios in multiattribute evaluation (prime)-elicitation and decision procedures under incomplete information. IEEE Trans. Syst. Man Cybern. Part A **31**(6), 533–545 (2001)
31. Tehrani, A.F., Cheng, W., Dembczyński, K., Hüllermeier, E.: Learning monotone nonlinear models using the Choquet integral. Mach. Learn. **89**(1–2), 183–211 (2012)
32. Wang, T., Boutilier, C.: Incremental utility elicitation with the minimax regret decision criterion, pp. 309–316 (2003)
33. Weng, P., Zanuttini, B.: Interactive value iteration for Markov decision processes with unknown rewards. In: Proceedings of IJCAI 2013, pp. 2415–2421 (2013)
34. White, C.C., Sage, A.P., Dozono, S.: A model of multiattribute decisionmaking and trade-off weight determination under uncertainty. IEEE Trans. Syst. Man Cybern. **14**(2), 223–229 (1984)
35. Yager, R.R.: On ordered weighted averaging aggregation operators in multicriteria decision making. IEEE Trans. Syst. Man Cybern. **18**(1), 183–190 (1988)

# Robustness of Approval-Based Multiwinner Voting Rules

Grzegorz Gawron[1,2] and Piotr Faliszewski[2(✉)]

[1] VirtusLab, Krakow, Poland
ggawron@virtuslab.com
[2] AGH University, Krakow, Poland
faliszew@agh.edu.pl

**Abstract.** We investigate how robust are approval-based multiwinner voting rules to small perturbations of the preference profiles. In particular, we consider the extent to which a committee can change after we add/remove/swap one approval, and we consider the computational complexity of deciding how many such operations are necessary to change the set of winning committees. We also consider the counting variants of our problems, which can be interpreted as computing the probability that the result of an election changes after a given number of random perturbations of the preference profile.

**Keywords:** Multiwinner voting · Approval voting · Robustness · Complexity

## 1 Introduction

The goal of a multiwinner election is to select a committee, i.e., a fixed-size set of candidates, based on the opinions of the voters. For example, citizens of a country may choose the members of their parliament, judges in a competition may select a group of its finalists, and a company may choose the products to offer based on the preferences of its customers; naturally, for each of these applications we would use a different voting rule, with different properties. Unfortunately, even the most appropriate rule may give unsatisfying results if the input votes are distorted. Indeed, some votes may be recorded erroneously due to mistakes of the voters or due to the mistakes of the election officials (or their machinery). In either case, it is interesting to know the consequences of such distortions. To address this issue, recently Bredereck et al. [5] initiated the study of robustness of multiwinner elections to small changes in the input votes. We follow up on their ideas, but instead of considering ordinal elections, we focus on the approval-based ones. (In the approval setting each voter simply indicates which candidates he or she finds acceptable; in the ordinal case, the voters rank the candidates from the most to the least appealing one.)

**Robustness in Multiwinner Elections.** We are interested both in the extent to which a winning committee can change—in the worst case—due to a small

© Springer Nature Switzerland AG 2019
S. Pekeč and K. B. Venable (Eds.): ADT 2019, LNAI 11834, pp. 17–31, 2019.
https://doi.org/10.1007/978-3-030-31489-7_2

perturbation of a single vote, and in computing the number of such perturbations necessary to change the result of a specific, given election. Additionally, we are also interested in computing the probability that introducing a given number of (randomly selected) perturbations changes the result.

Regarding the first issue, we use the notion of the *robustness level* of a multiwinner rule. Bredereck at al. [5] defined it to be $\ell$ if it is guaranteed that after swapping a pair of adjacent candidates in a vote (in the ordinal election model) the winning committee remains the same, except that up to $\ell$ candidates may be replaced (and, indeed, there are cases when this happens). In the approval setting, instead of swapping adjacent candidates we consider:

1. adding a single approval for a single candidate,
2. removing a single approval from a single candidate, or
3. swapping a single approval in a single vote (i.e., moving it from one candidate to the other).

In consequence, we consider three robustness level notions, one for each way of modifying the election.

Regarding the number of perturbations needed to change the election result, we consider three variants of the ROBUSTNESS RADIUS problem, originally introduced by Bredereck et al. [5], where we ask how many approvals have to be added, removed, or swapped to change the election result. To compute the probability that a number of randomly introduced perturbations change the election result, we consider counting versions of these problems (see below).

**Multiwinner Voting Rules.** We obtain our results by considering the following four voting rules: Approval Voting (AV), Satisfaction Approval Voting [4] (SAV), approval-based Chamberlin–Courant [9,26] (CCAV), and Proportional Approval Voting [30] (PAV). All these rules are different in their nature and capture different types of multiwinner elections. For example, AV chooses individually excellent candidates and can be used to select finalists of competitions, CCAV chooses diverse committees and can be used to select products that a company should offer to its customers [12,16], whereas PAV aims at representing the voters proportionally [1,21] and is well-suited for parliamentary elections [6]. SAV is closest in spirit to AV, but has its own quirks. We refer to the chapters of Kilgour [20] and Faliszewski et al. [14] for more details on multiwinner voting.

**Our Results.** As in the case of Bredereck et al. [5], we find that all our rules either have robustness level 1 (so that a single small change to the voters' preferences leads to replacing at most one candidate) or $k$ (so that a single change can lead to replacing the whole committee). Yet, we find two interesting quirks regarding the SAV rule. First, we find that even though the rule is polynomial-time computable, its robustness level—for the case of adding or deleting approvals—is $k$; this is the first example of such a simple rule with this behavior. The second quirk regarding SAV is that even though its robustness level for adding and removing approvals is $k$, for the case of swapping approvals it is 1.

**Table 1.** Summary of our results. In the three columns marked "Robustness Level" we provide the robustness level values for our rules; in the columns marked "Robustness Radius" we indicated the complexity results for the ROBUSTNESS RADIUS problem (including its counting variant for the AV and SAV rules; these results regard adding and removing approvals only, but not swapping them). We also indicate which of the rules are polynomial-time computable and which are NP-hard (but these results are not due to this paper).

| | | Robustness level | | | Robustness Radius (decision/counting) | | |
|---|---|---|---|---|---|---|---|
| Rule | Winner det | Adding approvals | Removing approvals | Swapping approvals | Adding approvals | Removing approvals | Swapping approvals |
| AV | P | 1 | 1 | 1 | P/FP | P/FP | P/— |
| SAV | P | $k$ | $k$ | 1 | P/#P-hard | P/#P-hard | P/— |
| CCAV | NP-hard | $k$ | $k$ | $k$ | NP-hard | NP-hard | NP-hard |
| | | | | | FPT($m$) | FPT($m$) | FPT($m$) |
| | | | | | FPT($n$) | FPT($n$) | FPT($n$) |
| PAV | NP-hard | $k$ | $k$ | $k$ | NP-hard | NP-hard | NP-hard |
| | | | | | FPT($m$) | FPT($m$) | FPT($m$) |
| | | | | | FPT($n$) | FPT($n$) | FPT($n$) |

Regarding the ROBUSTNESS RADIUS problem, we find that its decision variants are polynomial-time computable for AV and SAV, but become NP-hard for CCAV and PAV. Yet, for CCAV and PAV we find simple FPT algorithms parametrized by the number of candidates and by the number of voters. Interestingly, our FPT algorithm for the approval-based Chamberlin–Courant rule (for the parametrization by the number of voters) is much simpler than the analogous algorithm of Bredereck et al. [5] for the case of ordinal elections (as they did not consider PAV, we cannot compare the algorithms for this rule).

For AV and SAV, for which the ROBUSTNESS RADIUS problems are in P, we also consider their counting variants (for adding and removing approvals; we decided not to consider approval swapping for the sake of simplicity). The idea is as follows (we focus on adding approvals here): Let $X$ be the number of ways in which it is possible to add $B$ approvals to an election without changing the result, and let $Y$ be the total number of ways in which it is possible to add $B$ approvals. The fraction $X/Y$ gives the probability of *not changing the election result* by adding $B$ approvals uniformly at random. By computing this probability, we can distinguish situations where perturbing the election *may* possibly affect the election results but it is very unlikely, from those where we expect the result to be changed. It turns out that our counting problems are in FP for the AV rule, but are #P-hard for the SAV rule.[1]

---

[1] The class FP contains polynomial-time computable functions and the class #P is the analogue of NP for counting problems.

We summarize our results in Table 1. We omit many parts of the proofs due to limited space; in such cases, we will not repeatedly mention that the omission is due to space restriction, but we will simply state what we omit.

**Related Work.** Our work is most closely related to that of Bredereck et al. [5], which we have already discussed above. Two other closely related papers are due to Misra and Sonar [24] and Faliszewski et al. [17]. In the former, Misra and Sonar [24] consider the ROBUSTNESS RADIUS problem for the case of the Chamberlin–Courant rule, for some variants of single-peaked and single-crossing elections and for the approval setting. Their result for the case of approval-based Chamberlin–Courant rule is similar to ours and, to some extent, is stronger (in particular, they require fewer approved candidates per voter and, after dropping this restriction, they also obtain W[2]-hardness for the parametrization by the committee size), but the advantage of our result is that in a single proof we cover a large subfamily of Thiele rules [21,28], including CCAV and PAV.

In the latter paper, Faliszewski et al. [17] study the problem of bribery in approval-based election, where the goal is to ensure that a particular, specified candidate becomes a member of the winning committee by either adding, removing, or swapping approvals. Our work is similar to theirs in that we use the same types of operations, but our problems focus on changing the set of winning committees and not on ensuring some candidate's victory.

Prior to the work of Bredereck et al. [5], Shiryaev et al. [27] studied the robustness of single-winner elections. Specifically, they asked for the complexity of the DESTRUCTIVE SWAP BRIBERY problem, where the goal is to change the winner of an election by making as few swaps of adjacent candidates as possible (they considered the ordinal setting). Kaczmarczyk and Faliszewski [19] also studied a variant of this problem, where each swap involves a specified candidate (the original winner of the election). Other DESTRUCTIVE BRIBERY problems [13] were studied under the name MARGIN OF VICTORY [8,11,23,33] and focused on finding the smallest number of votes that need to be modified to change the election result. What makes our problem different is the focus on multi-winner approval elections (rather than ordinal ones) and looking for any change to the winning committee set (rather than having a specific candidate win).

Regarding the counting variant of the ROBUSTNESS RADIUS problem, so far relatively few authors considered counting variants of election-related problems. Two notable exceptions include the works of Hazon et al. [18] and Wojtas et al. [32], both focused on single-winner elections (however these certainly are not the only ones). In the former, the authors considered the problem of computing the probability that a given candidate is an election winner, provided that for each voter there is a probability distribution over the votes he or she may cast. In the latter, the authors considered the problem of computing the probability that a given candidate is a winner, provided that a given number of candidates or voters is added/removed from the election.

Finally, we mention that election robustness is a broad term that is studied in other contexts as well, e.g., in electronic voting and political science, but we omit the discussion of this literature as it is quite distant from our work.

## 2 Preliminaries

We assume basic familiarity with classic and parametrized complexity theory, including the notions of NP-hardness and fixed-parametrized tractability (FPT). We point to the textbooks of Papadimitriou [25] and Cygan et al. [10] for more details on these topics. For an integer $t$, we write $[t]$ to mean the set $\{1, \ldots, t\}$.

**Elections and Rules.** An *approval-based election* $E = (C, V)$ consists of a set of candidates $C = \{c_1, \ldots, c_m\}$ and a collection of voters $V = (v_1, \ldots, v_n)$, where each voter $v_i \in V$ approves a subset $C_{v_i}$ of candidates from $C$. For a voter $v_i$, we write $|v_i|$ to mean the number of candidates approved by this voter (i.e. $|v_i| = |C_{v_i}|$). For a candidate $c_j$, we write $V(c_j)$ to mean the set of voters that approve $c$, and we write $\text{app}_E(c_j)$ to mean the cardinality of this set. We sometimes refer to $\text{app}_E(c_j)$ as the approval score of candidate $c_j$.

An *approval-based multiwinner voting rule* $\mathcal{R}$ is a function that given an approval-based election and an integer $k \in [|C|]$ provides a family $\mathcal{R}(E, k)$ of size-$k$ committees, i.e., a family of size-$k$ subsets of $C$, that tie for victory. (We disregard the issue of tie-breaking and we consider all the provided committees as winning; in practical applications some tie-breaking rule would, of course, be necessary.)

Below we present several approval-based rules. Let $E = (C, V)$ be an election and fix committee size $k$. For each of the rules we define the score that it assigns to a given committee $S$; each rule selects the committees with the highest score:

**Approval Voting (AV).** Under the AV rule, the score of committee $S$ is defined as the sum of the approval scores of its members; formally we have:

$$\text{score}_E^{\text{AV}}(S) = \sum_{c \in S} \text{app}_E(c).$$

**Satisfaction Approval Voting (SAV).** Under the SAV rule, introduced by Brams and Kilgour [4], each voter has a single point which he or she distributes among all the candidates that he or she approves. For a candidate $c$, we refer to the value $\sum_{v \in V(c)} \frac{1}{|v|}$ as the SAV score of $c$. The SAV score of committee $S$ is the sum of the SAV scores of its members:

$$\text{score}_E^{\text{SAV}}(S) = \sum_{c \in S} \left( \sum_{v \in V(c)} \frac{1}{|v|} \right).$$

**Thiele Rules (Including CCAV and PAV).** A Thiele rule for the case of size-$k$ committees is defined through a vector $\omega = (\omega_1, \ldots, \omega_k)$ of non-negative real numbers. The score of committee $S$ under a Thiele rule specified by such a vector is defined as:

$$\text{score}_E^{\omega\text{-AV}}(S) = \sum_{v \in V} \left( \sum_{i=1}^{|C_v \cap S|} \omega_i \right).$$

Examples of Thiele rules include the AV rule, defined using vectors of the form $\omega^{\text{AV}} = (1, \ldots, 1)$, the approval-based Chamberlin–Courant rule (CCAV) [9,26], defined using vectors of the form $\omega^{\text{CCAV}} = (1, 0, \ldots, 0)$, and the Proportional Approval Voting rule (PAV), defined using vectors of the form $\omega^{\text{PAV}} = (1, 1/2, \ldots, 1/k)$. This family of rules was introduced by Thiele [30], who in particular introduced the PAV rule. We are mostly interested in Thiele rules with polynomial-time computable vectors $\omega = (\omega_1, \omega_2, \ldots, \omega_k)$, such that $1 = \omega_1 > \omega_2 \geq \omega_3 \geq \cdots \geq \omega_k$. We refer to such rules as *unit-decreasing Thiele rules*. CCAV and PAV are unit-decreasing, but AV is not.

Let us discuss CCAV, PAV, and Thiele rules in some more detail. Intuitively, under the CCAV rule a voter assigns score 1 to a committee if he or she approves at least one member of this committee, and assigns score 0 otherwise (so a voter is satisfied with a committee if it contains a candidate that the voter can see as his or her *representative*). Under the PAV rule, the appreciation of a voter for a committee increases with the number of its members that he or she approves, but the more members a voter approves, the smaller is the increase of his or her appreciation for the committee. The form of the vector used by PAV makes the rule suitable for parliamentary elections [1] and, indeed, it can be seen as a generalization of the D'Hondt apportionment method used in parliamentary elections in many countries [6].

There are polynomial-time algorithms for computing the winning committees under the AV and SAV rules [2], but the problem of deciding if there is a committee with at least a given score is NP-hard for CCAV [22,26], PAV [2,28], and many other Thiele rules (but there is a number of algorithmic workarounds, including approximation algorithms [7,29] and FPT algorithms [3,15]).

**Robustness Notions.** We adapt the definitions of Bredereck et al. [5] to the setting of approval voting. By the ADD operation, we mean adding an approval for a given candidate in a given vote, by the REMOVE operation we mean removing an approval from some candidate in a given vote, and by the SWAP operation we mean moving an approval from one candidate to another in a given vote.

For each operation Op $\in \{$ADD, REMOVE, SWAP$\}$ we say that a multiwinner rule $\mathcal{R}$ is $\ell$-Op-Robust if performing a single operation of type Op, in a given election, leads to replacing at most $\ell$ members of the winning committee. Formally, we have the following definition (the robustness level notion used by Bredereck et al. [5] was the same, except that they considered ordinal elections and the operation of swapping adjacent candidates in a preference order).

**Definition 1.** *Let $\mathcal{R}$ be a voting rule and let* Op $\in \{$ADD, REMOVE, SWAP$\}$ *be an operation type. We say that the* Op-*robustness level of $\mathcal{R}$ is $\ell$ ($\mathcal{R}$ is $\ell$-Op-robust) if $\ell$ is the smallest number such that for each election $E = (C, V)$ each committee size $k$, $k \leq |C|$, and each election $E'$ obtained from $E$ by applying a single* Op *operation the following holds: For each committee $W \in \mathcal{R}(E, k)$ there exists a committee $W' \in \mathcal{R}(E', k)$ such that $|W \cap W'| \geq k - \ell$.*

By a slight abuse of notation, we will often speak of $k$-Op-Robustness to mean that a single Op operation may lead to replacing the whole committee. We are also interested in Bredereck et al.'s [5] ROBUSTNESS RADIUS problem.

**Definition 2.** *Let $\mathcal{R}$ be a voting rule and let* OP $\in \{$ADD, REMOVE, SWAP$\}$ *be an operation type. In the $\mathcal{R}$-OP-ROBUSTNESS-RADIUS problem we are given an* *approval election $E = (C, V)$, a committee size $k$, an integer $B$, and we ask if by applying a sequence of $B$ operations of type OP it is possible to obtain an election $E'$ such that $\mathcal{R}(E, k) \neq \mathcal{R}(E', k)$.*

In addition to the above problems, we also consider their counting variants (for the ADD and REMOVE operations), where we consider the number of ways in which $B$ operations of a given type can be performed so that the result of the election *does not change*.[2]

## 3    Robustness Levels

In this section we analyze the robustness levels of our rules. We first consider the AV rule, which is 1-robust for each operation type.

**Proposition 1.** *AV rule is* 1-*Op-Robust for each* Op $\in$ {ADD, REMOVE, SWAP}.

The case of SAV is more intricate as the rule is $k$-ADD-robust and $k$-REMOVE-robust, but 1-SWAP-robust. Intuitively, adding or removing a single approval can affect the scores of many candidates so, in consequence, it can lead to replacing the whole committee. On the other hand, swapping an approval affects the scores of at most two candidates.

**Proposition 2.** *SAV is $k$-ADD-robust, $k$-REMOVE-robust, and 1-SWAP-robust.*

*Proof.* We consider the case of adding an approval only. Let us fix some committee size $k$ and let $A = \{a_1, \dots, a_k\}$ and $B = \{b_1, \dots, b_k\}$ be two disjoint sets of candidates. We form an election with candidate set $C = A \cup B$ and two voters, $v_1$ and $v_2$, such that $v_1$ approves all the candidates from $A$ and $v_2$ approves all the candidates from $B$.

Each candidate has SAV score $1/k$ and, in particular, $A$ is one of the winning committees. Yet, if we add an approval for $b_1$ to the vote of $v_1$, then the scores of all the candidates in $A$ decrease to $1/k+1$, whereas the scores of the candidates in $B$ remain unchanged (or increase, as in the case of $b_1$). In consequence, $B$ becomes the unique winning SAV committee. This completes the proof.    □

For the case of unit-decreasing Thiele rules (including CCAV and PAV), we find that they all are $k$-robust for each operation type. Our proof uses a single profile (parametrized by the committee size) for all the considered rules.

**Proposition 3.** *Every unit-decreasing Thiele rule is $k$-Op-robust for each* Op $\in$ {ADD, REMOVE, SWAP}.

---

[2] Since it is easy to compute the total number of ways of performing $B$ operations of a given type, from the computational complexity point of view it is irrelevant if we count the cases where the result changes or does not change , but the latter approach simplifies our proofs.

# 4   Complexity of the Robustness Radius Problems

In this section we focus on the complexity of the $\mathcal{R}$-OP-ROBUSTNESS-RADIUS problems for adding, removing, and swapping approvals. For the rules where our decision problems are polynomial-time solvable, we also consider the complexity of the respective counting problems. For the rules where the decision problems are NP-hard, we seek FPT algorithms.

## 4.1   The AV Rule: Polynomial-Time Algorithms

We start by considering the AV rule. In this case we have polynomial-time algorithms for all our decision and counting problems.

**Theorem 1.** AV-OP-ROBUSTNESS-RADIUS *is in* P *for each* OP $\in$ {ADD, REMOVE, SWAP}.

**Theorem 2.** *The problem of counting the number of ways in which B approvals can be added (removed) without changing the set of* AV *winning committees is in FP.*

*Proof.* We focus on the case of adding approvals (the case of removing approvals is similar). Let $E = (C, V)$ be an approval-based election, where $C = \{c_1, \ldots, c_m\}$ and $V = (v_1, \ldots, v_n)$, and let $k$ be the committee size. For each candidate $c_i$, let $z_i = \mathrm{app}_E(c_i)$ be its approval score. Without loss of generality, we assume that $z_1 \geq z_2 \geq \cdots \geq z_m$. Our goal is to count the number of ways in which it is possible to add $B$ approvals to the election so that the result does not change.[3] We consider two cases, either there is a unique winning committee in $E$ or there are several winning committees.

*Single Winning Committee.* There is a unique winning committee, $W = \{c_1, \ldots, c_k\}$, in election $E$ exactly if $z_k > z_{k+1}$. We need to count the number of ways of adding $B$ approvals so that afterwards each member of $W$ has higher approval score than each candidate outside of $W$. To this end, for each $\ell, u \in [n]$ we define the following two values:

1. $f(\ell, u)$ is the number of ways of adding $B$ approvals so that each member of $W$ has approval score at least $\ell$ and each candidate outside of $W$ has approval score at most $u$.
2. $g(\ell, u)$ is defined analogously to $f(\ell, u)$, except that we also require that at least one member of $W$ has approval score equal to $\ell$.

Our algorithm should output the value $\sum_{\ell=0}^{n} \sum_{u < \ell} g(\ell, u)$ as it covers all possibilities of adding $B$ approvals so that each member of $W$ has higher score than each candidate outside of $W$ without double-counting. To compute the values $g(\ell, u)$ in polynomial-time, we proceed as follows. First, we note that for each $\ell, u \in [n]$, we have that $g(\ell, u) = f(\ell, u) - f(\ell + 1, u)$.[4] This is so because

---

[3] We assume that it is, indeed, possible to add $B$ approvals, i.e., $z_1 + \cdots + z_m \leq nm - B$.

[4] For each $u \in [n]$, we take $f(n + 1, u)$ to be equal to $f(n, u)$ in this formula.

subtracting $f(\ell + 1, u)$ removes from consideration all the cases where all the members of $W$ have more than $\ell$ approvals. Second, we note that values $f(\ell, u)$ can be computed using dynamic programming (we omit the proof).

*Several Winning Committees.* We omit the proof due to limited space.    □

## 4.2   The SAV Rule: Easy Decision Problems, Hard Counting Ones

Let us now move on to the case of the SAV rule. The decision variants of our three problems are still polynomial-time computable for this rule.

**Theorem 3.** SAV-Op-Robustness-Radius *is in* $P$ *For each* Op $\in$ {Add, Remove, Swap}.

*Proof.* We consider the case of removing approvals only. Let $E = (C, V)$ be an input election, $k$ be the committee size, and $B$ be the number of operations that we can perform. We first consider the case that there is a unique SAV winning committee $X$ (we omit the case of multiple winning committees).

Let $Y = C \setminus X$ be the set of candidates that do not belong to $X$. Since $X$ is the unique winning committee, each of its members has a strictly higher SAV score than each of the candidates in $Y$. To change the election result, we must remove up to $B$ approvals so that there are two candidates, $x \in X$ and $y \in Y$, such that $y$ has at least as high SAV score as $x$ (indeed, if this were not the case then the winning committee certainly would not change, and if it is the case then either there is some winning committee that contains $y$ or there is some winning committee that does not contain $x$; in either case, the election result changes). Thus it suffices to try all pairs of candidates $x \in X$ and $y \in Y$ and test if by removing $B$ approvals we can ensure that $y$'s SAV score is greater or equal to that of $x$.

Fix candidate $x \in X$ and candidate $y \in Y$. Let $\Delta(y, x)$ mean the difference between the SAV score of $y$ and the SAV score of $x$ (in the original election we have $\Delta(y, x) < 0$; we aim to make it non-negative). We partition our election into four voter groups: (1) the voters who approve neither $x$ nor $y$, (2) the voters who approve $y$ but not $x$, (3) the voters who approve $x$ but not $y$, (4) the voters who approve both $x$ and $y$.

Note that there is no point in removing approvals from voters in Group 1, and for voters in Group 3 one should remove approvals for $x$ only. Our algorithm proceeds as follows. First, we guess three non-negative integers, $B_2$, $B_3$, and $B_4$, such that $B_2 + B_3 + B_4 = B$. Then we execute the following steps: (a) We choose $B_4$ voters from Group 4 who approve the fewest candidates and remove their approvals for $x$ (these voters now join Group 2); (b) We find $B_3$ voters from Group 3 who approve as few voters as possible and remove the approval for $x$ from each of them; (c) We execute the following operation $B_2$ times: We find a voter from Group 2 that approves the fewest candidates (but not only $y$) and remove an approval from him or her (but not regarding $y$).

If at any point during this algorithm there are not enough voters in a given group to perform a given operation, we disregard this guess of $B_2$, $B_3$, and $B_4$.

We accept if for some guess we reach a stage where $\Delta(y, x) \geq 0$ and we reject if this never happens. We omit the simple, but somewhat tedious, argument that our strategy is correct.  □

In contrast to the case of the AV rule, for SAV the problems of counting the number of ways in which $B$ approvals can be added or removed without changing the election result are #P-hard. We show this by giving a Turing reduction from the classic #P-hard problem, #PERFECT-MATCHINGS [31].

**Theorem 4.** *The problem of counting the number of ways in which approvals can be added (removed) without changing the set of SAV winning committees is #P-hard.*

*Proof.* We consider adding approvals only. We give a reduction from #PERFECT-MATCHINGS, where we are given a bipartite graph $G$ and we ask how many perfect matchings it has. We write $V(G)$ and $U(G)$ to denote the two sets of vertices in the graph, and we write $E(G)$ to denote its set of edges (so each edge from $E(G)$ connects vertex from $V(G)$ with some vertex from $U(G)$).

Let $n = |V(G)| = |U(G)|$ be the number of vertices in each part of the input graph (if $V(G)$ and $U(G)$ were of different cardinality, then there would be no perfect matching in our graph). We form an election $E = (C, V)$ as follows. We let the candidate set be $C = U(G) \cup V(G) \cup D$, where $D$ is a set of dummy candidates. Whenever we form a new voter who approves some dummy candidates, these dummy candidates are unique for this voter and are not approved by the other ones (thus, whenever we speak of a voter approving some dummy candidates, we implicitly create new dummy candidates). We create two groups of voters:

1. For each edge $e = \{u, v\} \in E(G)$, we form a voter $v_e$ who approves vertex candidates $u$ and $v$, and $n-2$ dummy candidates (thus, this voter contributes $1/n$ points to the SAV scores of $u$ and $v$). We refer to the voters in this group as to the *edge voters*.

2. For each vertex candidate $x \in U(G) \cup V(G)$, we form sufficiently many voters who each approve $x$ and $n-1$ dummy candidates, so that the SAV score of $x$ is exactly 2 (we add at most $2n$ voters for each vertex candidate; note that the edge voters contribute at most 1 point to the SAV score of each vertex candidate). We refer to the voters in this group as to the *filler voters*.

We set the size of the committee to be $k = 1$. In our election, each vertex candidate has SAV score 2 and each dummy candidate has SAV score $1/n$. Thus the winning committees are exactly the singleton subsets of $V(G) \cup U(G)$.

We will show that the number of ways in which it is possible to add $n$ approvals to our election without changing the set of winning committees is of the form $f(G) \cdot M(G)$, where $M(G)$ is the number of perfect matchings of $G$ and $f(G)$ is an easily computable function. This suffices to show #P-hardness of our problem. The intuition is that we can add $n$ approvals without changing the set of winning committees if and only if these approvals are added to the edge voters that correspond to a perfect matching (this way we decrease the scores of all the vertex candidates equally). Let us now consider what happens after we add some $n$ approvals:

**Some vertex candidate obtains an additional approval.** Let us assume that some vertex candidate $x$ obtains an additional approval in some vote $v$ (either a filler vote or an edge vote). Further, let $i - 1$ be the number of other candidates that also obtain an additional approval in this vote. Thus, $x$'s score increases by $\frac{1}{n+i}$ due to the additional approval. By adding the remaining $n - i$ approvals, we can decrease $x$'s score by at most $\frac{n-i}{n(n+1)}$. This is so, because adding a single approval to a vote where $x$ is already approved decreases $x$'s score by at most $\frac{1}{n} - \frac{1}{n+1} = \frac{1}{n(n+1)}$ (this happens when we add a single approval to an edge or filler vote where $x$ is approved; adding two approvals to such a vote decreases the score by $\frac{1}{n} - \frac{1}{n+2} = \frac{1}{n(n+2)}$, which is less than $2 \cdot \frac{1}{n(n+1)}$; in general, adding some $j$ approvals decreases $x$'s score by a lesser value than adding single approvals to $j$ votes where $x$ has originally been approved). However, we have that:

$$\frac{1}{n+i} - (n-i) \cdot \frac{1}{n(n+1)} = \frac{n+i^2}{n(n+1)(n+i)} > 0.$$

This means that after adding approvals, $x$'s score is greater than 2. Yet, by adding $n$ approvals it is impossible to increase the scores of more than $n$ of the $2n$ vertex candidates. This means that if some vertex candidate obtains an additional approval, then the set of winning committees changes.

**Some dummy vertex obtains an additional approval in a filler vote.** If some dummy candidate obtains an additional approval in a filler vote, then the score of the single vertex candidate who is approved in this vote decreases.[5] The remaining $n - 1$ additional approvals can lead to decreasing the scores of at most $2(n - 1)$ vertex candidates (e.g., if we add approvals in the edge votes only) so there is at least one vertex candidate whose score does not decrease. In consequence, the set of winning committees changes.

**Approvals are added only for dummy candidates in the edge votes.** Adding an approval for a dummy candidate in an edge vote $v_e$ decreases the scores of the vertex candidates incident to $e$ by exactly $\frac{1}{n(n+1)}$. Thus the only possibility that we add $n$ approvals and the scores of all the vertex candidates remain the same (but decreased) is that we add approvals to dummy candidates in $n$ edge votes that form a perfect matching (simple counting arguments show that if we add two approvals to some edge vote or if the edge votes do not form a perfect matching, then some vertex candidate's score does not decrease).

Let $M$ be the number of perfect matchings in our graph. By the above reasoning, for each perfect matching we have exactly $(|D| - (n-2))^n$ ways to add $n$ approvals in our election so that the set of winning committees does not change (for each edge from the matching, in the corresponding vote we can add an approval for one of the $|D| - (n - 2)$ dummy candidates that are not approved there). Thus, given a solution for our problem, it is easy to obtain the number of perfect matchings in the input graph.    $\square$

---

[5] From the previous case we know that it is possible to increase the score of this vertex candidate to the original value by adding some approval for him or her.

## 4.3    Unit-Decreasing Thiele Rules: Hardness and FPT Algorithms

For the case of unit-decreasing Thiele rules, all our decision problems are NP-hard (and, so, there is no point in considering their counting variants), but we obtain FPT algorithms for the parametrizations by the number of candidates and the number of voters. We start with the NP-hardness result. All the results in this section apply to CCAV and PAV.

**Theorem 5.** *Let* $\mathcal{R}$ *be a unit-decreasing Thiele rule and let* $\mathrm{Op}$ *be an operation in* $\{\mathrm{ADD}, \mathrm{REMOVE}, \mathrm{SWAP}\}$. $\mathcal{R}$-$\mathrm{OP}$-$\mathrm{ROBUSTNESS}$-$\mathrm{RADIUS}$ *is NP-hard.*

*Proof.* We consider the case of the SWAP operation only and we give a reduction from the classic NP-complete problem EXACT-COVER-BY-3-SETS (X3C). Our input X3C instance consists of a set $\mathcal{E} = \{e_1, ..., e_{3k}\}$ of $3k$ elements and a family $\mathcal{S} = \{S_1, ..., S_m\}$ of 3-element subsets of $\mathcal{E}$. The question is if there exists a family $\mathcal{S}^{cov} \subseteq \mathcal{S}$ of exactly $k$ sets such that each $e \in \mathcal{E}$ belongs to exactly one set in $\mathcal{S}^{cov}$.

Let us consider committees of size $k$ and let $\omega = (1, \alpha, \omega_3, \ldots, \omega_k)$ be the vector that specifies the unit-decreasing Thiele rule $\mathcal{R}$ for this setting. By definition, we have $1 > \alpha$. Let $\ell$ be the constant $\lceil \frac{3}{1-\alpha} \rceil$. We form election $E = (C, V)$ as follows. Let $A = \{a_1, ..., a_m\}$ and $B = \{b_1, ..., b_k\}$ be two sets of candidates, where the candidates in $A$ correspond to the sets from $\mathcal{S}$, and the candidates from $B$ form a default winning committee. We let the candidate set of our election be $C = A \cup B$ and we have the following voter groups (see Fig. 1):

1. We introduce voters $v_1, \ldots, v_{3k}$ who correspond to the elements from the set $\mathcal{E}$ (we refer to them as the element voters). Each voter $v_i$ approves candidate $b_{\lceil i/3 \rceil}$ and exactly those candidates $a_j$ for whom it holds that element $e_i$ belongs to the set $S_j$.
2. We introduce the set of voters $W = \{w_{i,j,l} \mid i \in [m], j \in [k], l \in [\ell]\}$, whose role is to form a mutual exclusion gadget regarding the winning committees (see details below). For each $i \in [m], j \in [k], l \in [\ell]$, $w_{i,j,l}$ approves $a_i$ and $b_j$.
3. We introduce the set of voters $W^a = \{w^a_{i,j,l} \mid i \in [m], j \in [m-k], l \in [\ell]\}$, whose role is to balance the scores of candidates in $A$ and $B$. For each $i \in [m], j \in [m-k], l \in [\ell]$, $w^a_{i,j,l}$ approves $a_i$.
4. We introduce voters $z_1, z_2$ who we use to control which committee wins. Both voters in $Z$ approve $b_1$.

We show that $B$ is the unique winning committee in election $E$. To this end, we first note that members of a given size-$k$ committee $S$ can receive at most $3k$ approvals from the element voters, $km\ell$ approvals from the voters in the sets $W \cup W^a$, and 2 approvals from the voters in $Z$. Thus, a size-$k$ committee $S$ must have $\mathrm{score}^{\omega\text{-}\mathrm{AV}}_E(S) \leq 3k + km\ell + 2$ (and for the equality to hold, it must be the case that each voter approves exactly one committee member). Since one can verify that $\mathrm{score}^{\omega\text{-}\mathrm{AV}}_E(B) = 3k + km\ell + 2$, $B$ is a winning committee. Furthermore, $B$ is a unique winning committee. To see this, let $Y$ be some winning committee different from $B$. Note that $Y$ must include candidate $b_1$ as this is the only way to get points from the voters in $Z$. For the sake of contradiction, suppose that

$Y$ also contains some candidate $a_i \in A$. However, this means that there are $\ell$ voters in $W$ who approve both $a_i$ and $b_1$. By definition of the unit-decreasing Thiele rules, the score that committee $Y$ receives from the $2\ell$ approvals that these voters grant to $a_i$ ad $b_1$ is at most $\ell + \alpha\ell < 2\ell$ and, so, the total score of committee $Y$ is lower than $3k + mk\ell + 2$. Thus, we have that $\mathcal{R}(E, k) = \{B\}$.

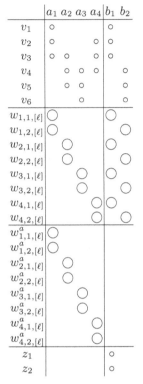

| | $a_1$ $a_2$ $a_3$ $a_4$ | $b_1$ $b_2$ |
|---|---|---|
| $v_1$ | ○ | ○ |
| $v_2$ | ○  ○ | ○ |
| $v_3$ | ○ ○  ○ | ○ |
| $v_4$ | ○ ○ ○ | ○ |
| $v_5$ | ○ ○ | ○ |
| $v_6$ | ○ | ○ |
| $w_{1,1,[\ell]}$ | ◯ | ◯ |
| $w_{1,2,[\ell]}$ | ◯ | ◯ |
| $w_{2,1,[\ell]}$ | ◯ | ◯ |
| $w_{2,2,[\ell]}$ | ◯ | ◯ |
| $w_{3,1,[\ell]}$ | ◯ | ◯ |
| $w_{3,2,[\ell]}$ | ◯ | ◯ |
| $w_{4,1,[\ell]}$ | ◯◯ | |
| $w_{4,2,[\ell]}$ | ◯ ◯ | |
| $w^a_{1,1,[\ell]}$ | ◯ | |
| $w^a_{1,2,[\ell]}$ | ◯ | |
| $w^a_{2,1,[\ell]}$ | ◯ | |
| $w^a_{2,2,[\ell]}$ | ◯ | |
| $w^a_{3,1,[\ell]}$ | ◯ | |
| $w^a_{3,2,[\ell]}$ | ◯ | |
| $w^a_{4,1,[\ell]}$ | ◯ | |
| $w^a_{4,2,[\ell]}$ | ◯ | |
| $z_1$ | | ○ |
| $z_2$ | | ○ |

**Fig. 1.** Example of an election used in the proof of Theorem 5, for X3C instance six elements $\{e_1, \ldots, e_6\}$ and sets $S_1 = \{e_1, e_2, e_3\}$, $S_2 = \{e_3, e_4, e_5\}$, $S_3 = \{e_4, e_5, e_6\}$, and $S_4 = \{e_2, e_3, e_4\}$. Symbol ○ represents an approval for a candidate from a given voter, and ◯ represents $\ell$ approvals coming from a group of voters.

We now show that it is possible to change the set of winning committees with a single approval swap if and only if the answer for our input X3C instance is *yes*.

($\longrightarrow$) Suppose that an exact cover $\mathcal{S}^{cov}$ exists for our input X3C instance and let $X$ be a size-$k$ committee corresponding to $\mathcal{S}^{cov}$ (i.e., $X$ contains members of $A$ that correspond to the sets in $\mathcal{S}^{cov}$).

We note that $\text{score}_E^{\omega\text{-AV}}(X) = 3k + km\ell$. To see this, note that as $X$ corresponds to a set cover, it receives $3k$ points from the element voters, each of the $k$ candidates in $X$ is approved by $k\ell$ voters from $W$, and is approved by $(m - k)\ell$ voters from $W^a$; each of these approvals translates to a single point because the voters in these groups do not approve more than one member of $X$ each. Finally, there are no approvals for members of $X$ from the voters in $Z$.

We form election $E'$ by swapping $z_2$'s approval from $b_1$ to some member of $X$, so we have $\text{score}_{E'}^{\omega\text{-AV}}(X) = 3k + km\ell + 1$. One can also verify that $\text{score}_{E'}^{\omega\text{-AV}}(b) = 3k + km\ell + 1$. As a consequence (and following similar reasoning as a few paragraphs above) we see that both $X$ and $A$ are winning committees. In other words, a single approval swap sufficed to change the set of winning committees.

($\longleftarrow$) We omit the details of the other direction. □

Fortunately, there are FPT algorithms for our problems for the parametrizations by the numbers of candidates and voters. The algorithms are inspired by those of Faliszewski et al. [17], but require some technical twists.

**Theorem 6.** *Let $\mathcal{R}$ be a unit-decreasing Thiele rule and let* OP *be an operation in* {ADD, REMOVE, SWAP}. *There are* FPT *algorithms for* $\mathcal{R}$-OP-ROBUSTNESS-RADIUS *both for the parametrization by the number of candidates and for the parametrization by the number of voters.*

## 5   Conclusion and Further Work

We have adapted the robustness framework of Bredereck et al. [5] to the approval setting. We have shown that the robustness levels of our rules are either 1 or $k$ (so small perturbations of the votes are either guaranteed to have minor effect only, or may completely change the results), and we have studied the complexity of deciding if a given number of perturbations can change election results.

**Acknowledgements.** This work was partially supported by the AGH University and the "Doktorat Wdrozeniowy" program of the Polish Ministry of Science and Higher Education.

## References

1. Aziz, H., Brill, M., Conitzer, V., Elkind, E., Freeman, R., Walsh, T.: Justified representation in approval-based committee voting. Soc. Choice Welfare **48**(2), 461–485 (2017)
2. Aziz, H., Gaspers, S., Gudmundsson, J., Mackenzie, S., Mattei, N., Walsh, T.: Computational aspects of multi-winner approval voting. In: Proceedings of AAMAS 2015, pp. 107–115 (2015)
3. Betzler, N., Slinko, A., Uhlmann, J.: On the computation of fully proportional representation. J. Artif. Intell. Res. **47**, 475–519 (2013)
4. Brams, S., Kilgour, D.: Satisfaction approval voting. In: Fara, R., Leech, D., Salles, M. (eds.) Voting Power and Procedures. Springer, Cham (2014). https://doi.org/10.1007/978-3-319-05158-1_18
5. Bredereck, R., Faliszewski, P., Kaczmarczyk, A., Niedermeier, R., Skowron, P., Talmon, N.: Robustness among multiwinner voting rules. In: Bilò, V., Flammini, M. (eds.) SAGT 2017. LNCS, vol. 10504, pp. 80–92. Springer, Cham (2017). https://doi.org/10.1007/978-3-319-66700-3_7
6. Brill, M., Laslier, J., Skowron, P.: Multiwinner approval rules as apportionment methods. J. Theor. Polit. **30**(3), 358–382 (2018)
7. Byrka, J., Skowron, P., Sornat, K.: Proportional approval voting, harmonic $k$-median, and negative association. In: Proceedings of ICALP 2018, pp. 26:1–26:14 (2018)
8. Cary, D.: Estimating the margin of victory for instant-runoff voting. In: Presented at 2011 electronic voting technology workshop/workshop on trushworthy elections, August 2011
9. Chamberlin, B., Courant, P.: Representative deliberations and representative decisions: proportional representation and the Borda rule. Am. Polit. Sci. Rev. **77**(3), 718–733 (1983)
10. Cygan, M., et al.: Parameterized Algorithms. Springer, Cham (2015). https://doi.org/10.1007/978-3-319-21275-3
11. Dey, P., Narahari, Y.: Estimating the margin of victory of an election using sampling. In: Proceedings of IJCAI-2015, pp. 1120–1126 (2015)
12. Elkind, E., Faliszewski, P., Skowron, P., Slinko, A.: Properties of multiwinner voting rules. Soc. Choice Welfare **48**(3), 599–632 (2017)
13. Faliszewski, P., Hemaspaandra, E., Hemaspaandra, L., Rothe, J.: Llull and Copeland voting computationally resist bribery and constructive control. J. Artif. Intell. Res. **35**, 275–341 (2009)

14. Faliszewski, P., Skowron, P., Slinko, A., Talmon, N.: Multiwinner voting: a new challenge for social choice theory. In: Endriss, U. (ed.) Trends in Computational Social Choice. AI Access Foundation (2017)
15. Faliszewski, P., Skowron, P., Slinko, A., Talmon, N.: Multiwinner analogues of the plurality rule: axiomatic and algorithmic views. Soc. Choice Welfare **51**(3), 513–550 (2018)
16. Faliszewski, P., Skowron, P., Slinko, A., Talmon, N.: Committee scoring rules: axiomatic characterization and hierarchy. ACM Trans. Econ. Comput. **6**(1), Article 3 (2019)
17. Faliszewski, P., Skowron, P., Talmon, N.: Bribery as a measure of candidate success: complexity results for approval-based multiwinner rules. In: Proceedings of AAMAS 2017, pp. 6–14 (2017)
18. Hazon, N., Aumann, Y., Kraus, S., Wooldridge, M.: On the evaluation of election outcomes under uncertainty. Artif. Intell. **189**, 1–18 (2012)
19. Kaczmarczyk, A., Faliszewski, P.: Algorithms for destructive shift bribery. In: Proceedings of AAMAS 2016, pp. 305–313 (2016)
20. Kilgour, M.: Approval balloting for multi-winner elections. In: Laslier, J.F., Sanver, M. (eds.) Handbook on Approval Voting. Springer, Heidelberg (2010). https://doi.org/10.1007/978-3-642-02839-7_6
21. Lackner, M., Skowron, P.: Consistent approval-based multi-winner rules. In: Proceedings of EC 2018, pp. 47–48 (2018)
22. Lu, T., Boutilier, C.: Budgeted social choice: from consensus to personalized decision making. In: Proceedings of IJCAI 2011, pp. 280–286 (2011)
23. Magrino, T., Rivest, R., Shen, E., Wagner, D.: Computing the margin of victory in IRV elections. In: Presented at 2011 electronic voting technology workshop/workshop on trushworthy elections, August 2011
24. Misra, N., Sonar, C.: Robustness radius for Chamberlin-Courant on restricted domains. In: Proceedings of SOFSEM 2019, pp. 341–353 (2019)
25. Papadimitriou, C.: Computational Complexity. Addison-Wesley, Reading (1994)
26. Procaccia, A., Rosenschein, J., Zohar, A.: On the complexity of achieving proportional representation. Soc. Choice Welfare **30**(3), 353–362 (2008)
27. Shiryaev, D., Yu, L., Elkind, E.: On elections with robust winners. In: Proceedings of AAMAS 2013, pp. 415–422 (2013)
28. Skowron, P., Faliszewski, P., Lang, J.: Finding a collective set of items: from proportional multirepresentation to group recommendation. Artif. Intell. **241**, 191–216 (2016)
29. Skowron, P., Faliszewski, P., Slinko, A.: Achieving fully proportional representation: approximability result. Artif. Intell. **222**, 67–103 (2015)
30. Thiele, T.: Om flerfoldsvalg. In: Oversigt over det Kongelige Danske Videnskabernes Selskabs Forhandlinger, pp. 415–441 (1895)
31. Valiant, L.: The complexity of computing the permanent. Theoret. Comput. Sci. **8**(2), 189–201 (1979)
32. Wojtas, K., Faliszewski, P.: Possible winners in noisy elections. In: Proceedings of AAAI 2012, pp. 1499–1505, July 2012
33. Xia, L.: Computing the margin of victory for various voting rules. In: Proceedings of EC 2012, pp. 982–999. ACM Press, June 2012

# Towards Characterizing the Deterministic Combinatorial Constrained Efficient Space

Rica Gonen[1(✉)] and Anat Lerner[2]

[1] Department of Management and Economics, The Open University of Israel,
The Dorothy de Rothschild Campus, 1 University Road, 43107 Raanana, Israel
gonenr@openu.ac.il
[2] Department of Mathematics and Computer Science, The Open University of Israel,
The Dorothy de Rothschild Campus, 1 University Road, 43107 Raanana, Israel
anat@cs.opeu.ac.il

**Abstract.** It is often the case that economic mechanisms can have hidden characteristics with unexpected consequences in practice. This is especially true when real-world budgets come into play. For instance, the infamous German and British 3G spectrum auctions that generated revenue so successfully that cellular service providers delayed the rollout of 3G networks for lack of funds.

We contribute a piece of the puzzle by characterizing the space of deterministic, dominant-strategy incentive compatible, individually rational, and Pareto-optimal combinatorial auctions where efficiency does not hold. We examine a model with two players and $k$ nonidentical items ($2^k$ outcomes), multidimensional types, private values, nonnegative prices, and quasilinear preferences for the players with one relaxation - one of the players is subject to a publicly-known budget constraint.

We show that if it is publicly known that the valuation for the largest bundle is more than the budget for at least one of the players then the following are true. (a) VCG does not fulfill the basic properties of deterministic, dominant-strategy incentive compatible, individual rationality and Pareto optimality when the all-item bundle is not arbitrarily allocated. (b) Of the dictatorial solutions, only a single family of non-trivial dictatorial mechanisms fulfills the above basic properties. (c) Weakening the public knowledge assumption results in no VCG nor dictatorship mechanisms that fulfill the properties. Our characterization of the non-efficient space for deterministic budget-constrained combinatorial auctions is similar in spirit to that of [20] for Bayesian single-item constrained efficiency auctions.

**Keywords:** Budget constraints · Dictatorship · Pareto efficiency · Incentive compatibility

© Springer Nature Switzerland AG 2019
S. Pekeč and K. B. Venable (Eds.): ADT 2019, LNAI 11834, pp. 32–48, 2019.
https://doi.org/10.1007/978-3-030-31489-7_3

# 1   Introduction

The inclusion of budgets and heterogeneity when designing economic mechanisms better characterizes many real world problems such as commonly studied bandwidth (combinatorial) auctions. However, this inclusion greatly complicates design and the outcomes of such mechanisms can be difficult to predict. Consider the unexpected result of the German and British 3G radio spectrum auctions in early 2000. In these auctions telecom companies overspent to the point of depleting the funds needed to roll out 3G networks, resulting in a delayed 3G rollout. However, a change in the auction mechanism or even the market conditions could have created the opposite effect of companies paying nothing for spectrum rights or a single company acquiring all of the spectrum rights. These types of outcomes are not limited to spectrum auctions and could easily occur in an automated globalized supply chain. The drive for lean globalized supply chains has substantially increased competition among suppliers and driven down margins. As such, some suppliers try to win business though they are incapable of funding the contracted quantity/quality of procured goods, which introduces the confounding factor of players' budgets to the problem of characterizing the outcome of practical auctions.

We contribute a piece of the overall puzzle by characterizing the space of deterministic, dominant-strategy incentive compatible (IC), individually rational, and Pareto-optimal combinatorial auctions where efficiency does not hold. We examine a model with two players and $k$ nonidentical items ($2^k$ outcomes), multidimensional types, private values, nonnegative prices, and quasilinear preferences for the players with one relaxation - one of the players is subject to a publicly-known budget constraint. This setting is somewhat more complex than that of common auction literature as it adds budgets and heterogeneity, which more accurately describe mechanisms used in practice.

The phenomena in the German and British 3G radio spectrum auctions as well as the present day proliferation of suppliers highlights the potential gap between willingness to pay and ability to pay, and the potential of better understanding how budget constraints affect auction outcomes. Further consider that most goods are not sold in uniform bundles or used independent of other items. Though blocks of radio bandwidth are apparently uniform they are not identical, as well can be said for goods in supply chain auctions which are bundled to fulfill diverse bills of materials. The addition of the seemingly minor dimension of heterogeneity profoundly affects auction design complexity.

Our result characterizes a space where efficient outcomes do not exist in this context. For instance, the characterization answers whether and under which conditions it is possible for a small telecommunications company to compete meaningfully with better financed companies in a bandwidth auction.

[20] showed that in a Bayesian setting where one indivisible item is for sale there exists a threshold value such that if the players have a valuation that is more than the threshold then the solution is inefficient, Bayesian IC, and individually rational in expectation. [18] showed for the two item case, that there exists a unique family of dictatorial solutions in deterministic combinatorial auctions

that are dominant-strategy IC, individually rational, and Pareto-optimal with publicly-known budget-constraints and publicly known high valuations. Moreover it was shown [11] that if trivial dictatorial mechanisms, i.e., dictatorial mechanisms with zero revenue, are ruled out by a natural anonymity property then an impossibility of design is revealed and there is no deterministic combinatorial auction that is dominant-strategy IC, individually rational, and Pareto optimal where players have publicly-known budget constraints and the all-item bundle is nonarbitrarily allocated (a property termed nonarbitrary hoarding), in a model with multiple nonidentical items and nonnegative prices. Therefore, some additional public knowledge needs to be assumed to allow a large space of mechanisms to exist other than the trivial dictatorial mechanism. Indeed [19] characterized the combinatorial efficiency space in the aforementioned setting assuming that it is a public knowledge that at least one of the players value the $k$-item bundle less than the constrained player's budget. [19] proved that VCG [6,13,24] is the only mechanism that fulfills the above properties.

More specifically, in this paper we assume the negation of the public knowledge assumed in [19] and prove that no VCG mechanism exists under this assumption. Nevertheless we show that if it is publicly known that at least one of the players values the $k$-item bundle more than the constrained player's budget, the player has non zero value for all other bundles and the allocation of the $k$-item bundle is nonarbitrary, i.e., no trivial dictatorial mechanism is considered[1], then there exists a *single* family of dictatorial mechanisms that is dominant-strategy IC, individually rational and Pareto optimal. Furthermore any weakening of our suggested public assumption results in a space with no dictatorial mechanisms at all in addition to no VCG mechanisms. In other words we prove the minimal public knowledge assumption under which (non-trivial) dictatorial mechanisms that fulfill the above desired properties exist and the VCG mechanism does not exist. Our dictatorial mechanisms are a natural extension of Arrow's definition of a dictatorial social welfare function to a monetary domain and differs from Groves dictatorial mechanisms as negative prices are not included in our model.

There are classic as well as recent results showing that dictatorship (or sequential dictatorship) is the only mechanism that is not subject to individual manipulations and is Pareto optimal in mechanism design models without assuming quasilinearity (See e.g. [1,10,23]) Arrow's seminal impossibility [1] shows that for unrestricted domains (at least three possible outcomes) under determinism and transitivity axioms, independence of irrelevant alternatives (IIA), and Pareto-optimality conditions, every social choice function must be a dictatorship or imposed. However, the conditions of Arrow's theorem as well as the conditions of [10,23] can be satisfied when the requirement for unrestricted domains is relaxed, as was shown for one dimensional domains such as single peaked. While the possibility/impossibility of maintaining Arrow's desired properties is known for the whole space of the nonmonetary domain of preferences, much is yet left to be understood when restricting attention to the assumption of side payments and transferable currency.

---

[1] The formal definition of the nonarbitrary allocation is defer to Sect. 2.

Throughout the paper we assume deterministic mechanisms. To understand the role of determinism in our result one must look into the literature of nondeterministic constrained auctions such as [20]. [20]'s work defines the properties of constrained-efficient auctions, *i.e.*, maximizing the expected social welfare under Bayesian incentive compatibility and budget-constrained players. [20] states that the domains of efficiency and inefficiency are determined by a threshold value. The computation of the threshold value makes use of expectation and allows for allocations with negative utility for the players. Therefore [20]'s threshold cannot be used in an individually rational deterministic setting. The domains of inefficiency that can be concluded from our analysis are determined by the budget. The immediate implication of the budget as the threshold of inefficiency is that the budget cannot be a privately known value but must be publicly known. As such, in our deterministic setting the budget is publicly known much like [7,9,16], and [11]; while in [4,20]'s nondeterministic setting the budgets are privately known.

### 1.1   Our Contribution

As the complication introduced by budgets and heterogenous goods makes characterizing the space necessarily complex, this section presents a summary of the result and its implications, informed speculation on possible inferences and a concrete example demonstrating the base result. In summery we show that if it is publicly known that the valuation for the largest bundle is more than the budget for at least one of the players then the following are true. (a) VCG does not fulfill the basic properties of deterministic, dominant-strategy IC, individual rationality and Pareto optimality when the all-item bundle is not arbitrarily allocated. (b) Of the dictatorial solutions, only a single family of non-trivial dictatorial mechanisms fulfills the above basic properties. (c) Weakening the public knowledge assumption results in no VCG nor dictatorship mechanisms that fulfill the properties. Our characterization of the non efficient space for deterministic budget-constrained combinatorial auctions is similar in spirit to that of [20] for Bayesian single-item constrained efficiency auctions.

In general terms we show that in our setting one of the following outcomes will arise. If the value of the budget-limited player is more than his own budget, either the player with the unlimited budget (practically speaking) will be allocated all of the goods for the budget of the limited player or the players will share the goods and both will potentially pay nothing. On the other hand, if the player with an unlimited budget values the goods more than the budget-limited player is able to pay then the players share the goods and both will potentially pay nothing.

We also speculate that the general space of dominant-strategy IC combinatorial auctions with budgets most likely includes only VCG and dictatorial mechanisms. Also, it appears that the dictatorial aspect depends on the introduction of budgets. We draw the above conclusion as the research community has indications to believe that the general space of dominant-strategy IC combinatorial auctions without budgets includes only VCG mechanisms. Therefore, the discovery of dictatorial mechanisms in our study is most likely brought about by our inclusion of players with budgets.

As an example illustrating this result, consider two telecommunications companies $T_1$ and $T_2$ that are competing for bandwidth in a region with $k$ distinct broadcast bands. Company $T_1$ has a limited budget $b$ to spend on bandwidth. On the other hand company $T_2$ is a large telecommunications company with practically unlimited funds to spend on the discussed region. If company $T_1$ values taking control of all the bandwidth available in the region more than $b$ then $T_2$'s bandwidth allocation will be determined and priced such that it receives its most beneficial bandwidth allocation potentially for free, regardless of any high values stated by company $T_1$. The applicability of the results is the indication to the auction designer under what circumstances the outcome might be less desirable, such as zero revenue or dict allocation if he wishes to maintain the common economic properties. Given that in practice some participants are budget constrained and most practical mechanism design problems call for good heterogeneity.

We explicitly study a model with two players. Indeed there may be a gap between the two player model and a model with more than two players, i.e., with more possible outcomes. However, the literature tends to indicate that an outcome space as large as the $2^k$ model's can capture the complexity of the outcome of the $n$ player $k$ items model (for any finite $n$ and $k$) of the multidimensional combinatorial auction possibility space. This level of applicability was indicated by Roberts' characterization of the complete preference domain space [22]. In Roberts' result a three-outcome model captures the complexity of the whole $n$-outcome possibility space of the preference multidimensional space for any finite $n$. Hence, we believe that extending our model to more players (where one of them is budget constrained) will maintain the unique dictatorial solution when it is publicly known that at least one of the players values the $k$-item bundle more than the constrained player's budget, has non zero value for all other bundles and the allocation of the $k$-item bundle is nonarbitrary. We leave the extension of our proof for future research.

## 1.2   Prior Literature

It is well known that in quasilinear environments with a complete preference domain over at least three outcomes and non-constrained players, only VCG mechanisms (and weighted VCG) ([6,13,24]) satisfy the dominant-strategy IC property [22][2]. Nevertheless when preferences are subject to free disposal and no externalities are assumed, as is common in combinatorial auctions, then the possibility space of dominant-strategy IC combinatorial auctions in the multidimensional type model is not yet defined. Several papers investigate computationally feasible dominant-strategy IC (but inefficient) auctions with one-dimensional private values e.g. [17] as well as with multidimensional private-value settings with some additional restrictions on the preference space [8].

---

[2] In quasilinear environments only Groves mechanisms satisfy the dominant-strategy IC and Pareto optimal properties ([12,14]).

In recent years several papers studied budget-constrained combinatorial auctions. [7] showed that there does not exist a deterministic auction that is individually rational, dominant-strategy IC, and Pareto optimal with potentially negative prices and privately known budgets, even when players are one-dimensional types. [9] showed that the same impossibility holds for one-dimensional types with different items and publicly known multi-item demand. [16] also showed the same impossibility with publicly known budgets if multidimensional types (two identical items with three outcomes) are considered.

[7, 9, 16] allow negative prices to exist. This means that some players are paid for participation in the auction, either by the mechanism or by the other players. Practical auction implementations usually can not afford or are unwilling to consider paying bidders for their participation nor are they interested in encouraging side payments among the participants. Therefore, similar to [20]'s and [11]'s model we chose to assume that all prices are nonnegative. This assumption narrows down the domain of possible allocations in comparison to the potential negative prices model with multidimensional types. Nevertheless, some of the mechanisms that fulfill the three properties of dominant strategy IC, individually rational, and Pareto optimal in the nonnegative price model are not included in the mechanism space that fulfills the same properties in the negative price model. The reason for the above derives from the Pareto optimality property. Since the model with nonnegative prices has a smaller set of possible allocations there exist situations where a mechanism does not fulfill the Pareto optimal property in the model with negative prices but does fulfill the Pareto optimal property in the model with nonnegative prices.

[7] also characterizes the possibility space of dominant-strategy incentive compatibility and Pareto optimal budget-constrained combinatorial auction mechanisms. [7]'s characterization is restricted to one-dimensional types and therefore their possibility space characterization does not imply the possibility space in our model with multidimensional types. More specifically, [7] showed that for multi-unit demand and identical items, Ausubel's clinching auction, which assumes public budgets and additive valuations, uniquely satisfies the properties described above. In a similar model with small randomized modification [4] showed that [7]'s result can be obtained with private budgets. Similarly Ausubel's clinching auction was concluded by [9] for one-dimensional types with different items and publicly known multi-item demand. For unit-demand players with private values and budget constraints in the one-dimensional types model there are several deterministic mechanisms that fulfill the properties of IC and Pareto optimality (see e.g. [2]). In nondeterministic mechanisms with a one-dimensional types model (one indivisible unit) [20] characterizes constrained-efficiency mechanisms, which are mechanisms that maximize the expected social welfare under Bayesian incentive compatibility and budget constraints in a nonnegative price model.

There are few other works that focus on revenue maximization under budget constraints in a single-item auctions settings (e.g. [3, 5, 15, 21]).

## 2   Notation and Definitions

We consider combinatorial auction mechanisms with $k$ different types of items and 2 players. Let $N = \{1, 2\}$ be the set of players and $C = \{c_1, \cdots, c_k\}$ be the set of items. Let $\mathscr{B}$ be the set of all subsets of items, that is, $\mathscr{B} = 2^C$.

Each player $i$ has a private value $v_i(B)$ for every bundle $B \in \mathscr{B}$ drawn from a valid valuation space $\mathscr{V}_i{}^3$, i.e., players are multi-minded and have different private values for different bundles of the items. We denote player $i$'s private values by a $2^k$-tuple: $V_i = \{v_i(B) | B \in \mathscr{B}\}$. We assume that $v_i(\emptyset) = 0$, that is, for both players the valuation of the empty bundle is zero. We also assume that $v_i(B') \leq v_i(B)$ whenever $B' \subseteq B$, meaning free disposal: for both players the allocation of an extra item can not reduce their valuation (the usual assumption in combinatorial auctions). As players are multi-minded and have different private values for different bundles of the items, a player $i$ may have a separate arbitrary value for each of the $2^k$ possible outcomes. This means our valuation space is a multidimensional valuation space and the players have multidimensional valuations.

We assume that player 1 has a limited budget $b_1$ while player 2 has an unlimited budget for acquiring the items and these budgets are publicly known information.

We denote the auction mechanism

$F(V_1, V_2, b_1) = (B_1, B_2, p(B_1), p(B_2))$ where $B_i$ is the bundle allocated to player $i$ and $p(B_i)$ is the price of the bundle $B_i$. We assume that all the prices are nonnegative, i.e., $p(B_i) \geq 0$ for $i = \{1, 2\}$. We also assume that the auction mechanism $F(V_1, V_2, b_1)$ can produce at least $2^k - 1$ outcomes, i.e., there exist at least $2^k - 1$ inputs with different outputs.

**Definition 1.** *The utilities of player 1, player 2 and the auctioneer are defined as follows:*
*Player 1's utility is*

$$u_1(F(V_1, V_2, b_1)) = \begin{cases} v_1(B_1) - p(B_1) \ if \ p(B_1) \leq b_1 \\ -\infty \qquad\qquad\qquad otherwise \end{cases}$$

*Player 2's utility is $u_2(F(V_1, V_2, , b_1)) = v_2(B_2) - p(B_2)$.*
*The auctioneer's utility is $u_a(F(V_1, V_2, b_1)) = p(B_1) + p(B_2)$.*

For simplicity of notation, whenever $F, V_1, V_2$ and $b_1$ are clear from the context we denote $u_i(F(V_1, V_2, b_1))$ by $u_i$.

**Definition 2. Allocation Space**
*We say that a mechanism is limited to an **allocation space** $AS = (B_1, B_2 = C - B_1, p(B_1), p(B_2))$, if the feasible outcome space is restricted to AS regardless of the valuations.*

---

[3] Along the paper we consider valuation spaces where not all valuations are included in the valuation space.

## Definition 3. Dictatorial

*An auction mechanism $F(V_1, V_2, b_1)$ is called a **dictatorial mechanism** if there exists a player $i \in \{1, 2\}$ (**the dictator**) such that given the allocation space $AS$, $F$ maximizes the utility of the dictator regardless of the other player's valuations. Formally, $F(V_1, V_2, b_1) = (B_1, B_2, p(B_1), p(B_2))$ is a dictatorial mechanism given $AS$, and player $i$ is the dictator if:*

$$\forall(V_1, V_2) \in (\mathscr{V}_1, \mathscr{V}_2), \quad F(V_1, V_2, b_1) \in AS \quad and$$
$$(\forall V_i \in \mathscr{V}_i, \ and \ \forall(B_1', B_2', p(B_1'), p(B_2')) \in AS),$$

$$(v_i(B_i) - p(B_i) \geq v_i(B_i') - p(B_i'))$$

Our definition of a dictatorial mechanism is a natural extension of Arrow's dictatorial social welfare function to a monetary domain. To understand the connection between our definition of a dictatorial mechanism and Arrow's dictatorial definition consider [1]'s definition: "A social welfare function is said to be "dictatorial" if there exists an individual $i$ such that for all $x$ and $y$, $xP_iy$ implies $xPy$ regardless of the ordering of all individuals other than $i$, where P is the social preference relation corresponding to those orderings." Arrow's definition implies that the dictator determines the outcome regardless of the other players' preferences similar to the way our dictator determines the outcome allocation regardless of the other player's preferences, though in a monetary domain. As our model is in the monetary domain our definition also adds the monetary aspect and implies that the dictator's payment is independent of the other player's preferences.

In the paper we will consider only dictatorship mechanisms that are Pareto Optimal (see definition 6) and therefore they will all be of the following form: The mechanism first maximizes the utility of the dictator and then the mechanism selects an allocation that maximizes the utility of the other player from the allocations the dictator is indifferent to.

We next define the four properties under which [11]'s impossibility holds: Individual Rationality (IR), Nonarbitrary Hoarding, Pareto Optimality and Truthfulness.

## Definition 4. Property 1: Individual Rationality (IR)

*An auction mechanism $F(V_1, V_2, b_1)$ is called **individually rational** if for every player $i$, $u_i(F(V_1, V_2, b_1)) \geq 0$. Specifically, the following must hold:*

1. $v_1(B_1) - p(B_1) \geq 0$ and $p(B_1) \leq b_1$ *(IR of Player 1)*
2. $v_2(B_2) - p(B_2) \geq 0$ *(IR of Player 2)*

Note that the auctioneer's utility is nonnegative from our assumption that all the prices are nonnegative.

## Definition 5. Property 4: Truthfulness

*An auction mechanism $F(V_1, V_2, b_1)$ is called **Truthful** if none of the two players can increase his own utility by reporting false valuations. That is, given the true valuations $(V_1, V_2) \in (\mathscr{V}_1 \times \mathscr{V}_2)$, for every $(V_1', V_2') \in (\mathscr{V}_1 \times \mathscr{V}_2)$ the following hold:*

- $u_1(F(V_1, V_2', b_1)) \geq u_1(F(V_1', V_2', b_1))$
- $u_2(F(V_1', V_2, b_1)) \geq u_2(F(V_1', V_2', b_1))$

### Definition 6. Property 3: Pareto Optimality (PO)

*An auction mechanism $F$ is called* **Pareto optimal** *if for every input $(V_1, V_2, b_1)$, such that $(V_1, V_2)$ in $(\mathcal{V}_1 \times \mathcal{V}_2)$, there is no allocation $(B_1', B_2', p(B_1'), p(B_2'))$ such that all the following inequalities hold, with at least one strong inequality:*

- $u_1(B_1', p(B_1'), B_2', p(B_2')) \geq u_1(F(V_1, V_2, b_1))$

- $u_2(B_1', p(B_1'), B_2', p(B_2')) \geq u_2(F(V_1, V_2, b_1))$

- $u_a(B_1', p(B_1'), B_2', p(B_2')) \geq u_a(F(V_1, V_2, b_1))$

At first glance it may seem that our definition of PO may ask too much of the mechanism by including the auctioneer's utility. Several papers include the auctioneer's utility in the PO definition (e.g. [7]). We included the auctioneer's utility to allow a larger space of mechanisms to exist. In the case of dictatorial mechanisms if only the buyers' utility is included in the PO definition then any mechanism that does not give both players a price of exactly 0 is not PO.

We denote the following mechanism as a **trivial mechanism**. $F(V_1, V_2, b_1) = (B_1 = C - B_2, B_2, p(B_1) = p(B_2) = 0)$, s.t. each of the two players can be the dictator. The dictator is allocated the most valued bundle and the other player is allocated the most valued bundle from the dictator's indifference set and they both pay zero.

The trivial mechanism is the only dictatorial mechanism that satisfies the three properties of IR, Pareto optimality and truthfulness as was shown in [11]. The property of nonarbitrary hoarding that was first presented in [11] aims to eliminate the trivial mechanism. With the nonarbitrary hoarding property assumed there is no mechanism that satisfies the four properties of IR, Pareto optimality, truthfulness and nonarbitrary hoarding.

### Definition 7. Property 2: Nonarbitrary Hoarding

*Given $V = \{V_1, V_2\}$, a general set of private valuations in $\mathcal{V}_1 \times \mathcal{V}_2$; and $B_1$ and $B_2$, the output bundles of $F(V_1, V_2, b_1)$. An auction mechanism $F$ is called* **nonarbitrary hoarding** *if the following two conditions hold:*

1. *If $B_2 = \{c_1, \cdots, c_k\}$*
   *then $v_2(c_1, \cdots, c_k) \geq \min\{b_1, v_1(c_1, \cdots, c_k)\}$.*
2. *If $B_1 = \{c_1, \cdots, c_k\}$*
   *then $\min\{b_1, v_1(c_1, \cdots, c_k)\} \geq v_2(c_1, \cdots, c_k)$.*

*Intuitively, a mechanism fulfills nonarbitrary hoarding if whenever all $k$ items are allocated to a single player, the player is chosen nonarbitrarily, i.e., according to the valuations of the two players and the budget. Furthermore the player chosen has to be able to afford the $k$-item bundle more than the other player. Note that the property of nonarbitrary hoarding does not require if and only if. That is, a mechanism can maintain nonarbitrary hoarding even when $v_2(c_1, \cdots, c_k) \geq$*

$\min\{b_1, v_1(c_1, \cdots, c_k)\}$ *and player 2 is not allocated the k-item bundle, or when* $\min\{b_1, v_1(c_1, \cdots, c_k)\} \geq v_2(c_1, \cdots, c_k)$ *and player 1 is not allocated the k-item bundle.*

As discussed in Sect. 1 in order to circumvent [11] impossibility, some additional public knowledge needs to be assumed. We define four possible publicly known restrictions on the valuations space:

**R1** $v_1(c_1, \cdots, c_k) < b_1$
**R2** $v_2(c_1, \cdots, c_k) < b_1$
**R3** $(v_1(c_1, \cdots, c_k) \geq b_1)$ and $(\exists\, 0 < m \leq b_1,\ s.t.\ (\forall B \neq \emptyset,\ v_1(B) \geq m))$
**R4** $v_2(c_1, \cdots, c_k) > b_1$ and $(\exists\, 0 < m \leq b_1,\ s.t.\ (\forall B \neq \emptyset,\ v_2(B) \geq m))$

R3 and R4 means that the player is completely multi minded that is to say that the player has a non zero value for every bundle that is not the empty set.

## 3   Impossibility Space - No VCG Mechanisms

[19] proved that if one of restrictions R1 or R2 is publicly known then the only mechanism that satisfies the four properties is an efficient mechanism with VCG prices. In what follows we show that in the negation valuation space of R1 or R2 there does not exist a VCG mechanism that fulfills the four properties of truthfulness, IR, PO, and nonarbitrary hoarding.

CLAIM 1. *If we do not assume public knowledge of R1 or R2, as assumed in* [19], *that is, we do not assume that* $v_1(c_1, \cdots, c_k) < b_1$ *or* $v_2(c_1, \cdots, c_k) < b_1$, *then the unique VCG mechanism proven in* [19] *for the domain will not satisfy the three properties of: IR, PO, and truthfulness.*

PROOF OF CLAIM 1. *Suppose that both* $v_1(c_1, \cdots, c_k)$ *and* $v_2(c_1, \cdots, c_k)$ *are not restricted to be lower than* $b_1$. *Consider the following valuations for* $B \notin \emptyset$:
$v_2(c_1, \cdots, c_k) = b_1 + \epsilon$ *s.t.* $\epsilon < b_1/4$ *and* $v_2(B \neq (c_1, \cdots, c_k)) = b_1 - \epsilon$
  $v_1(c_1, \cdots, c_k) = 2b_1$ *and* $v_1(B, \neq (c_1, \cdots, c_k)) = b_1$
  *Then, the allocation* $B_1 = (c_1, \cdots, c_k)$ *with VCG prices, does not satisfy IR for player 1 as* $p(B_1 = (c_1, \cdots, c_k)) = v_2(c_1, \cdots, c_k) > b_1$ *and player 1 is budget constrained. Any allocation* $(B_1, C - B_1)$, *such that* $B_2 \neq (c_1, \cdots, c_k)$ *with VCG prices is not IR for player 2 as* $v_2(B_2) = b_1 - \epsilon$ *and the VCG price is* $v_1(c_1, \cdots, c_k) - v_1(C - B_2) = 2b_1 - b_1 = b_1 > v_2(B_2)$.
  *Therefore the only allocation is* $B_2 = (c_1, \cdots, c_k)$, *for the price* $\min\{b_1, v_1(c_1, \cdots, c_k)\} = b_1$.
  *But this allocation is not PO as any allocation of non-empty bundles for the two players* $(B_1, C - B_1)$ *with the prices:* $p(B_1) = b_1 - \epsilon$ *and* $p(B_2) = 2\epsilon$, *will be better for both players and the auctioneer.*

## 4  Possibility Space - Dictatorial Mechanisms

In this section we study the valuation spaces for which there exist dictatorial mechanisms that satisfy the four properties of IR, Pareto optimality, truthfulness and nonarbitrary hoarding. In the next two subsections we investigate the effect of public knowledge of one player's restricted valuation space. In Subsect. 4.1 we assume that the fact that $\mathcal{V}_1$ satisfies restriction R3 is publicly known and prove that of the dictatorial solutions, only a single family of non-trivial dictatorial mechanisms fulfills the above basic properties. In the presented family the only possible dictator is player 2 and the family has $2^k$ feasible outcomes. Furthermore we prove that R3 is the minimal restriction on $\mathcal{V}_1$ that allows nontrivial mechanism solutions.

In Subsect. 4.2 we assume that it is publicly known that $\mathcal{V}_2$ satisfies restriction R4 and prove that of the dictatorial solutions, only a single family of non-trivial dictatorial mechanisms fulfills the above basic properties. In the presented family the only possible dictator is player 1 and the family has $2^k - 1$ feasible outcomes.

### 4.1  When R3 Is Publicly Known

In this subsection we study the effect of R3 as public knowledge, i.e., $(v_1(c_1, \cdots , c_k) \geq b_1)$ and $(\exists\, 0 < m \leq b_1,\ s.t.\ (\forall B \neq \emptyset,\ v_1(B) \geq m))$. The main theorem of this subsection follows.

**Theorem 1.** *Given that restriction R3 is publicly known:*

1. *Of the dictatorial solutions there exists a single family of nontrivial dictatorial mechanisms that satisfies the four properties of IR, Pareto optimality, truthfulness and nonarbitrary hoarding.*
2. *The family is unique up to a price parameter $y$ and a valuation parameter $m > 0$, $0 \leq y \leq m \leq b_1$, such that $\forall B_1 \neq \emptyset$, $p(B_1) = y$.*
3. *Player 2 is the sole dictator.*
4. *$p(B_2 = \{c_1, \cdots , c_k\}) = b_1$, and $p(B_2 \neq \{c_1, \cdots , c_k\}) = 0$.*
5. *For any weaker restriction on player 1's valuations, there is no nontrivial dictatorial mechanism that satisfies the properties of IR, Pareto optimality, truthfulness and nonarbitrary hoarding.*

PROOF OF THEOREM 1. *We prove the theorem with the help of the following lemmas:*

– *Lemma 1 proves that in any dictatorial mechanism that satisfies the four properties, all the $2^k - 2$ nonempty partitions of $C$ between two players must be feasible outcomes.*
– *Lemma 2 proves that if R3 is publicly known then in addition the two allocations of the k-item bundle to each player must also be feasible outcomes.*
– *Lemma 3 proves that if R3 is publicly known, then player 1 can not be the dictator.*

- *Lemma 4 proves the uniqueness of the prices. I.e., for any dictatorial mechanism with $2^k$ outcomes that does not apply the following price structure at least one of the properties is violated. $y$ is a parameter, s.t., $m \geq y \geq 0$.*

$$p(B_1) = \begin{cases} 0 \; If \; B_1 = \emptyset \\ y \; Otherwise \end{cases}$$

$$p(B_2) = \begin{cases} b_1 \; If \; B_2 = \{c_1, \cdots, c_k\} \\ 0 \; Otherwise \end{cases}$$

- *Lemma 5 proves that given that restriction R3 is publicly known, the dictatorial mechanism with $2^k$ feasible outcomes that applies the price structure defined in Lemma 4 and in which player 2 is the dictator, satisfies all of the four properties.*
- *Lemma 6 proves that restriction R3 is the minimal restriction on player 1's valuations that results in a mechanism with strictly positive prices that satisfies the four properties.*

**Lemma 1.** *In any dictatorship that satisfies the properties of IR, Pareto optimality, truthfulness and nonarbitrary hoarding, all of the $2^k - 2$ nonempty partitions of $C$ between two players must be feasible outcomes.*

*The proof of Lemma 1 is omitted due to space limitations.*

**Lemma 2.** *Given that restriction R3 is publicly known, in any dictatorial mechanism that satisfies the properties the outcomes $(B_1 = \{c_1, \cdots, c_k\}, B_2 = \emptyset, p(B_1), p(B_2) = 0)$ and $(B_1' = \emptyset, B_2' = \{c_1, \cdots, c_k\}, p(B_1') = 0, p(B_2'))$ must also be feasible.*

*The proof of Lemma 2 is omitted due to space limitations. In the following lemma we study the identity of the dictator.*

**Lemma 3.** *If R3 is publicly known then Player 1 can not be the dictator.*

PROOF OF LEMMA 3.1. *Suppose to the contrary that R3 is publicly known and player 1 is the dictator. Consider the following situation.*
$\forall B \neq \{c_1, \cdots, c_k\}, \quad v_1(c_1, \cdots, c_k) - b_1 > v_1(B), \quad v_2(c_1, \cdots, c_k) > b_1$. *Then from player 1's IR we conclude that $p(B_1 = \{c_1, \cdots, c_k\}) \leq b_1$, and therefore player 1's utility from the k-item bundle, even for the highest price of $b_1$, is strictly higher than his utility from any other bundle even if this bundle is given for free. Therefore, as player 1 is the dictator, any dictatorial mechanism must allocate $B_1 = \{c_1, \cdots, c_k\}$. However, this allocation contradicts nonarbitrary hoarding as $\min\{b_1, v_1(c_1, \cdots, c_k)\} = b_1 < v_2(c_1, \cdots, c_k)$.*

*In the following lemma we define a price structure and prove that for any other price structure at least one of the properties is violated.*

**Lemma 4.** *Let $0 \leq y \leq m$ be a constant parameter and consider the following price structure:*

$$p(B_1) = \begin{cases} 0 \quad If \; B_1 = \emptyset \\ y \quad Otherwise \end{cases}$$

$$p(B_2) = \begin{cases} b_1 & If \ \ B_2 = \{c_1, \cdots, c_k\} \\ 0 & Otherwise \end{cases}$$

*Given that restriction R3 is publicly known, any dictatorial mechanism that applies a price structure different from the above price structure, violates at least one of the properties.*

*We prove Lemma 4 with the help of the following five claims. The proofs of the claims are omitted due to space limitations.*

CLAIM 2. *The price paid by player 2 for any given bundle must be a constant.*

CLAIM 3. *For every $i \in \{1, 2\}$, If $B_i \subseteq B_i'$ then $p(B_i) \leq p(B_i')$*

PROOF OF CLAIM-3.1. *This follows directly from truthfulness and free disposal.*

CLAIM 4. $p(B_2 = \{c_1, \cdots, c_k\}) = b_1$

CLAIM 5. $\forall B_2 \neq \{c_1, \cdots, c_k\}, \ \ p(B_2) = 0$

CLAIM 6. $\forall B_1 \neq \emptyset, \ \ m \geq p(B_1) = y \geq 0.$

*In the following lemma we prove that the dictatorship with $2^k$ outcomes with player 2 as the dictator and the price structure defined in Lemma 4 satisfies the four properties.*

**Lemma 5.** *The dictatorial mechanism defined in Theorem 1 satisfies the four properties of IR, truthfulness, Pareto optimality, and nonarbitrary hoarding.*

*The proof of Lemma 5 is omitted due to space limitations.*

*[19] showed that if there is no public knowledge of the players' valuations then there is no mechanism that satisfies the four properties of IR, truthfulness, Pareto optimality, and nonarbitrary hoarding. [19] further showed that if it is publicly known that $v_1(c_1, \cdots, c_k) < b_1$ then the only mechanism that satisfies the four properties is the efficient mechanism with VCG prices[4]. Therefore, if there is a restriction on $\mathcal{V}_1$ that supports a dictatorial mechanism it must be at least $v_1(c_1, \cdots, c_k) \geq b_1$.*

*The slightest weakening of R3 is therefore R3' $v_1(c_1, \cdots, c_k) \geq b_1$ and $\exists B_1 \neq \emptyset$, and $(\exists \ 0 < m \leq b_1, \ s.t. \ (\forall B \not\subseteq B_1 \ v_1(B) \geq m))$. Meaning that player 1's valuation for at least one bundle $B_1$ is not bounded from below.*

**Lemma 6.** *If R3' is publicly known then there is no dictatorial mechanism that satisfies the properties of IR, truthfulness, nonarbitrary hoarding and Pareto optimality.*

*The proof of Lemma 6 is omitted due to space limitations.*

---

[4] Note that from free disposal $v_1(c_1, \cdots, c_k) < b_1 \ \ \Rightarrow \ \ \forall B, \ v_1(B) < b_1.$

## 4.2   When R4 Is Publicly Known

In this subsection we study the effect of R4 as public knowledge, i.e., $v_2(c_1, \cdots, c_k) > b_1$ and $(\exists\, 0 < m \leq b_1,\ s.t.\ (\forall B \neq \emptyset,\ v_2(B) \geq m))$.

The main theorem in this subsection is Theorem 2:

**Theorem 2.** *Given that restriction R4 is publicly known:*

1. *Of the dictatorial solutions, there exists a single family of nontrivial dictatorial mechanisms that satisfies the four properties of IR, Pareto optimality, truthfulness and nonarbitrary hoarding, with $2^k - 1$ feasible outcomes. The only outcome that is not feasible is allocating an empty bundle to player 2.*
2. *Player 1 is the sole dictator.*
3. *The family is unique up to a price parameter $x$ and a valuation parameter $m > 0$, $0 \leq x \leq m \leq b_1$, such that $\forall B_2 \neq \emptyset,\ p(B_2) = x$.*
4. *$\forall B_1,\ p(B_1) = 0$.*

PROOF OF THEOREM 2. *We prove the theorem with the help of the following lemmas.*

- *Lemma 1 proved that in any dictatorial mechanism that satisfies the four properties, all the $2^k - 2$ nonempty partitions of $C$ between two players must be feasible outcomes.*
- *Lemma 7 proves that if R4 is publicly known then in addition the allocation of the k-item bundle to player 2 must also be feasible and the allocation of an empty bundle to player 2 can not be feasible.*
- *Lemma 8 proves that if R4 is publicly known, and player 2 is the dictator, then the only dictatorial mechanism that satisfies the four properties is the trivial mechanism.*
- *Lemma 9 proves that there exists a unique family of nontrivial dictatorial mechanisms with player 1 as the dictator that satisfies the four properties.*

**Lemma 7.** *Given that restriction R4 is publicly known in any dictatorial mechanism that satisfies the properties of IR, truthfulness, Pareto optimality, and nonarbitrary hoarding, the outcome $(B_1 = \emptyset, B_2 = \{c_1, \cdots, c_k\}, p(B_1) = 0, p(B_2))$ must also be feasible and the outcome $(B_1 = \{c_1, \cdots, c_k\}, B_2 = \emptyset, p(B_1), p(B_2) = 0)$ can not be feasible.*

PROOF OF LEMMA 7.1. *Suppose that the outcome $B_2 = \{c_1, \cdots, c_k\}$ is not feasible. Consider the valuations $\forall B \neq \{c_1, \cdots, c_k\}$,*

$$v_2(c_1, \cdots, c_k) > v_2(B),\ v_1(c_1, \cdots, c_k) = b_1 - \varepsilon,\ v_1(B \neq \{c_1, \cdots, c_k\}) = 0.$$

*Then from R4 we conclude that the allocation $B_2 = \emptyset$ contradicts nonarbitrary hoarding. Any other allocation $(B_1 \notin \{\emptyset, \{c_1, \cdots, c_k\}\},\ B_2 = C - B_1,\ p(B_1),\ p(B_2))$ is not Pareto optimal as allocating $B_2' = \{c_1, \cdots, c_k\}$ for $p(B_2)$ is strictly better for player 2 while the auctioneer and player 1 are indifferent to the two alternatives. The outcome $(B_1 = \{c_1, \cdots, c_k\}, B_2 = \emptyset, p(B_1), p(B_2) = 0)$ can not be feasible as it contradicts nonarbitrary hoarding since $v_2(c_1, \cdots, c_k) > b_1 \geq \min\{b_1, v_1(c_1, \cdots, c_k)\}$.*

*In the following lemma we prove that if R4 is publicly known the only dictatorial mechanism dictated by player 2 is the trivial mechanism, where* $p(B_1) = p(B_2) = 0$.

**Lemma 8.** *Given that restriction R4 is publicly known and that player 2 is the dictator, the only dictatorial mechanism that satisfies the properties of IR, truthfulness, Pareto optimality, and nonarbitrary hoarding is the trivial mechanism.*

*We prove Lemma 8 with the help of the following five claims. The proofs of the claims are omitted due to space limitations.*

CLAIM 7. $0 \le p(B_2 = \{c_1, \cdots, c_k\}) = x \le b_1$.

CLAIM 8. *If R4 is publicly known then* $\forall B_2$, $B_2' \notin \{\{c_1, \cdots, c_k\}, \emptyset\}$ $p(B_2) = p(B_2')$.

CLAIM 9. $\forall B_2 \ne \{c_1, \cdots, c_k\}$, $p(B_2) = 0$.

CLAIM 10. *If R4 is publicly known, and player 2 is the dictator then* $\forall B_1$, $p(B_1) = 0$.

*To complete the proof of Lemma 8 we claim that if R4 is publicly known then* $x = 0$, *i.e., any dictatorial mechanism must be the trivial mechanism with* $p(B_1) = p(B_2) = 0$.

CLAIM 11. *If R4 is publicly known then* $x = 0$.

**Lemma 9.** *Given that R4 is publicly known, there exists a unique family of nontrivial dictatorial mechanisms with player 1 as the dictator that satisfies the four properties.*

*The proof of Lemma 9 is omitted due to space limitations.*

## 5   Concluding Remarks

Our result takes a step toward showing that the characterization of deterministic budget-constrained combinatorial auctions is structurally similar to Maskin's seminal work characterizing the Bayesian budget-constrained single-item auctions. Maskin showed that if the players have a valuation above a certain threshold then the solution is inefficient (lottery among dictators), Bayesian, IC, and IR in expectation. [19] proved that if R1 or R2 is publicly known then the only mechanism that satisfies the four properties is an efficient mechanism with VCG prices.

We contribute a piece of the puzzle by proving that in the negation of R1 or R2 (Maskin's equivalent to over the threshold) there does not exist an efficient mechanism with VCG prices that fulfills the four properties. We also prove that in the negation of R1 and R2 (i.e. R3 and R4) there are dictatorial mechanisms that fulfill the four properties. We also show the minimality of R3 (R3') in which no efficient or dictatorial mechanisms exist.

We speculate that the general space of dominant-strategy IC combinatorial auctions with budgets most likely includes only VCG and dictatorial mechanisms. It appears that the dictatorial aspect depends on the introduction of budgets. We draw the above conclusion as the research community has indications to believe that the general (multidimensional) space of dominant-strategy IC combinatorial auctions without budgets includes only VCG mechanisms. Therefore, the discovery of dictatorial mechanisms in our study is most likely brought about by our inclusion of players with budgets.

# References

1. Arrow, K.: A difficulty in the concept of social welfare. J. Polit. Econ. **58**(4), 328–346 (1950)
2. Ashlagi, I., Braverman, M., Hassidim, A., Lavi, R., Tennenholtz, M.: Position auctions with budgets: existence and uniqueness. B.E. J. Theor. Econ. (Advances) **10**(1), 20 (2010)
3. Benoit, J., Krishna, V.: Multiple object auctions with budget constrained bidders. Rev. Econ. Stud. **68**, 155–179 (2001)
4. Bhattacharya, S., Conitzer, V., Munagala, K., Xiax, L.: Incentive compatible budget elicitation in multi-unit auctions. In: The 21st Annual ACM-SIAM SODA, pp. 554–572 (2010)
5. Che, Y., Gale, I.: Standard auctions with financially constrained bidders. Rev. Econ. Stud. **65**, 1–21 (1998)
6. Clarke, E.: Multipart pricing of public goods. Public Choice **2**, 17–33 (1971)
7. Dobzinski, S., Lavi, R., Nisan, N.: Multi-unit auctions with budget limits. Games Econ. Behav. **74**, 486–503 (2008)
8. Dobzinski, S., Nisan, N.: Multi-unit auctions: beyond roberts. In: The 12th ACM Conference on Electronic Commerce EC, pp. 233–242 (2011)
9. Fiat, A., Leonardi, S., Saia, J., Sankowski, P.: Single valued combinatorial auctions with budgets. In: The 12th ACM Conference on Electronic Commerce EC, pp. 223–232 (2011)
10. Gibbard, A.: Manipulation of voting schemes: a general result. J. Econometrica. **41**, 587–601 (1973)
11. Gonen, R., Lerner, A.: The incompatibility of pareto optimality and incentive compatibility in quasilinear settings with anonymous and budget constrained players. Games **4**(4), 690–710 (2013)
12. Green, J., Laffont, J.: Characterization of satisfactory mechanisms for the revelation of preferences for public goods. J. Econometrica. **45**, 427–438 (1977)
13. Groves, T.: Incentives in teams. Econometrica **41**, 617–631 (1973)
14. Holmstrom, B.: Groves' scheme on restricted domains. J. Econometrica. **47**, 1137–1144 (1979)
15. Laffont, J., Robert, J.: Optimal auction with financially constrained buyers. Econ. Lett. **52**, 181–186 (1996)
16. Lavi, R., May, M.: A note on the incompatibility of strategy-proofness and pareto-optimality in quasi-linear settings with public budgets. Econ. Lett. **115**, 100–103 (2012)
17. Lehmann, D., O'Callaghan, L., Shoham, Y.: Truth revelation in approximately efficient combinatorial auctions. J. ACM **49**(5), 577–602 (2002)

18. Lerner, A., Gonen, R.: Dictatorial mechanisms in constrained combinatorial auctions. B.E. J. Theor. Econ. **13**(1), 1–18 (2013). https://doi.org/10.1515/bejte-2013-0006

19. Lerner, A., Gonen, R.: Characterizing the incentive compatible and pareto optimal efficiency space for two players, $k$ items, public budget and quasilinear utilities. Games **5**(2), 1–9 (2014)

20. Maskin, E.: Auctions, development and privatization: efficient auctions with liquidity-constrained buyers. Eur. Econ. Rev. **44**, 667–681 (2000)

21. Pai, M.M., Vohra, R.: Optimal auctions with financially constrained bidders. In: Working paper (2008)

22. Roberts, K.: The characterization of implementable choice rules. Aggregat. Revel. Pref. **12**(2), 321–348 (1979)

23. Satterthwaite, M.: Strategy-proofness and arrow's condition: existence and correspondence theorems for voting procedures and social welfare functions. J. Econ. Theor. **10**, 187–217 (1975)

24. Vickrey, W.: Counterspeculation, auctions and competitive sealed tenders. J. Finan. **16**, 8–37 (1961)

# Integrating Operators' Preferences into Decisions of Unmanned Aerial Vehicles: Multi-layer Decision Engine and Incremental Preference Elicitation

Arwa Khannoussi[1(✉)], Alexandru-Liviu Olteanu[1], Christophe Labreuche[2],
Pritesh Narayan[3], Catherine Dezan[1], Jean-Philippe Diguet[1],
Jacques Petit-Frère[1,2,3,4], and Patrick Meyer[1]

[1] Univ. Brest, IMT Atlantique, Université Bretagne Sud, CNRS, Lab-STICC,
UMR CNRS 6285, Brest, Lorient, France
{arwa.khannoussi,catherine.dezan}@univ-brest.fr,
{arwa.khannoussi,patrick.meyer}@imt-atlantique.fr,
{alexandru.olteanu,jean-philippe.diguet}@univ-ubs.fr
[2] Thales Research & Technology, Palaiseau Cedex, France
{christophe.labreuche,jacques.petit-frere}@thalesgroup.com
[3] University of the West of England, Bristol, UK
pritesh.narayan@uwe.ac.uk
[4] Thales Group, Elancourt Cedex, France

**Abstract.** Due to the nature of autonomous Unmanned Aerial Vehicles (UAV) missions, it is important that the decisions of a UAV stay consistent with the priorities of an operator, while at the same time allowing them to be easily audited and explained. We therefore propose a multi-layer decision engine that follows the logic of an operator and integrates its preferences through a Multi-Criteria Decision Aiding model. We also propose an incremental approach to elicit the operator's preferences, in view of minimizing his/her cognitive fatigue during this task.

**Keywords:** Incremental preference elicitation ·
Multi-layer decision engine · Multi-Criteria Decision Aiding ·
Operator's preferences · Traceable decisions · Autonomous UAV

## 1 Introduction

Autonomous Unmanned Aerial Vehicles (UAVs) are capable of carrying out various types of missions (military or civilian). Throughout the mission, they may face multiple choices and have to make many decisions without any human input. This decision making task requires that multiple, potentially conflicting, criteria are taken into account in order to achieve the mission's and the operator's objectives. In addition, in order to increase his/her confidence in the UAV's behavior, its decisions should be consistent with the priorities of the operator.

© Springer Nature Switzerland AG 2019
S. Pekeč and K. B. Venable (Eds.): ADT 2019, LNAI 11834, pp. 49–64, 2019.
https://doi.org/10.1007/978-3-030-31489-7_4

Previous research has focused primarily on the decision making of autonomous UAVs through the perspective of trajectory calculation taking into account different constraints and objectives, by using optimization techniques [2,4,13]. Recently, deep learning techniques have also been used to tackle decision problems of autonomous UAVs, as,for example, the path planning problem [7] or the selection of high level directives [14,21]. In [16] the authors suggest to integrate the operator's perspective into the calculation of trajectories for autonomous UAVs, and propose to use techniques from the field of Multi-Criteria Decision Aiding to model the operator's preferences. They ground their proposal on the hypothesis that an operator will trust the behavior of an autonomous UAV if it makes decisions that are consistent with his/her priorities.

We start from the same observation as [16], but propose to integrate the preferences of the operator even more deeply in the various decision making tasks of autonomous UAVs. We therefore develop in this contribution:

- a multi-layer decision engine for autonomous UAVs, which mimics the logic adopted by operators during a non-autonomous mission,
- the integration of a Multi-Criteria Decision Aiding (MCDA) model, called Simple Ranking Method using Reference Profiles (SRMP) [18], into this decision engine, which allows the autonomous UAV to select the appropriate high-level action to be executed during the mission,
- an incremental preference elicitation approach to tune the SRMP decision model according to the preferences of the operator, while minimizing his/her cognitive fatigue during the learning process.

We also validate the interest of our proposal through a simulator, in which we can test the influence of different operator profiles on the UAV's behaviour.

The article is structured in the following way. Section 2 provides a state of the art on existing work. Section 3 presents an overview of the proposed multi-layer decision making engine while a description of the considered preference model is given in Sect. 4, next to our proposal for an incremental preference elicitation procedure. A validation of our proposal, is then presented in Sect. 5, before finishing with some concluding remarks and perspectives in Sect. 6.

## 2   State of the Art

### 2.1   Decisions in Autonomous UAVs

From a general point of view, an established body of work exists which focuses on trajectory calculation while taking into account different constraints and objectives. Blackmore et al. [4] present an approach to calculate the optimal trajectory in the presence of obstacles and uncertain information, while Kabama et al. [13] illustrate an approach to calculate the optimal trajectory for combat UAVs by avoiding radars. Some approaches also address the calculation of the trajectory in a non-convex environment with uncertainties [2].

Delmerico et al. [7] propose the use of Convolutional Neural Networks (CNNs), for path planning in the context of collaborative search and rescue missions. Deep learning is also used to present a solution for UAV localization and cross-view localization of images in [22]. Concerning the UAV navigation task, some advances have led to the application of CNNs in order to map images to high-level behavior directives (e.g., turn left, turn right, rotate left, and rotate right) [14,21]. Due to resource limitations, the learned model is executed off-board, on the GPU of an external laptop. The work presented by Tipaldi and Glielmo [25] integrates Markovian Decision Processes (MDP) for spacecraft reconfiguration in order to deal with the uncertainty in the outcome of actions and is applied to autonomous mission planning and execution.

Using a drone with a high level of autonomy to perform a mission requires that the human operator has a high degree of confidence in the capacity of the drone to make the "right" decisions. This observation motivated Narayan et al. [16] to integrate preferences into the calculation of the objective function in order to generate trajectories that more accurately represent the preferences of an operator. For that they use a decision model from the field of Multi-Criteria Decision Aiding.

## 2.2  Multi-Criteria Decision Aiding and (incremental) Preference Elicitation

In this article, we start from the same observation as [16], but propose to integrate the operator's preferences into higher level decisions rather than the calculation of trajectories, as, e.g., the choice between decision actions that the drone has to perform, as landing, returning to the base, aborting the mission, skipping a waypoint, etc.

Multi-Criteria Decision Aiding (MCDA) [19] is the study of decision problems and methods which may be used in order to assist a decision maker (DM) in reaching a decision when faced with a set of so-called alternatives (or decision actions), described via multiple, often conflicting, criteria. Two main methodological schools have been proposed to support DMs facing a multi-criteria decision problem: outranking methods and multi-attribute value theory (MAVT). In our autonomous UAV context, the DM is the operator, whose preference model is integrated into the drone and is guiding its decisions.

The preference parameters of MCDA decision models can be given directly by the DM through a direct preference elicitation approach. However, such an approach is usually too difficult to implement in practice, as the DM needs to have a very good understanding of the MCDA model. Therefore, a second approach is to start from partial knowledge on the output of the method. This *indirect* preference elicitation has received much attention from researchers, as for example in the seminal works of Jacquet-Lagreze et al. [12] in the MAVT context and of Ngo The et al. [24] in the outranking context. These techniques generally determine in one shot the parameters configuration compatible with the input provided by the DM, and are therefore *not incremental* by nature.

*Incremental preference elicitation* focuses on learning the parameters of a decision model in a streaming setting. Incremental learning algorithms receive learning data sequentially, one by one or chunk by chunk, and use this data with the previously learned model to produce a new, better one, that encapsulates information held by the data seen so far. Regarding this progressiveness, in the MAVT context, Durbach [9] and Lahdelma et al. [15] use an index that quantifies the volume of the polyhedron of the constraints specifying the possible value functions. Holloway et al. [11] show the importance of the order of the pair-wise comparisons in decreasing the number of questions for reducing the cognitive effort of the DM. Ciomek et al. [6] present a set of heuristics to minimize the number of elicitation questions and prioritize them in the context of *single choice* decision problems. In the same context, Benabbou et al. [3] select a set of pair-wise questions using a minimax regret strategy.

An incremental elicitation of the parameters of MCDA models should reduce the cognitive effort of the DM, as he/she is facing only to a limited number of questions. As we will show in this article, the decision model that we are integrating into autonomous UAVs is learned incrementally before the mission, and it is important that the operator is not overly stressed during this phase.

## 3    Onboard Multi-layer Decision Engine

The starting point of our proposal is the hypothesis that an operator will trust the behavior of an autonomous UAV if it makes decisions, which are consistent with his/her priorities. Our proposal differs from more classical mission planning through its ability to react to unanticipated events and the integration of the operator's preferences in the decision engine of the drone. Furthermore, we also propose a decision model which can be easily explained and whose outcomes (decisions) are easily interpretable, so that the operator can validate the decision engine implemented in the UAV.

Consequently we focus on the logic adopted by the operator during a mission, in order to define the model of the autonomous decision engine. We first suppose that, in the context of manned aerial vehicles, the human operator does not make decisions continuously during the flight, but that the decision making act is triggered by events (e.g. the appearance of an obstacle, a breakdown, a significant change in weather conditions ... ). Second, still in a non-autonomous context, we also suppose that when operators have to deal with a complex decision, triggered by an event, they tend to decompose it into a sequence of sub-decisions. Consequently, in case of such an event, the operator will take into consideration possible trajectories (i.e. which is a sub-decision) while choosing a high level action (e.g. land, continue the mission, skip a waypoint, ... ).

Following this reasoning, we propose to decompose the decision-making process of the autonomous UAV during the mission into two layers (Fig. 1).

**Layer 1** is for monitoring the mission progress and all the information that might impact its success. Based on the occurrence of certain events, the second layer may be triggered. These events could be related to the UAV's environment (e.g. a change of the flight zone, the appearance of an obstacle, the detection of heavy rainfall, a mechanical breakdown, ... ) or to risk levels (e.g. exceeding a certain threshold of risk towards the drone, the mission, or the environment, ... ). While we do not tackle this topic in our current proposal, a series of rules or even a preference model, that is previously tuned to the perspective of an operator, may be integrated into the drone.

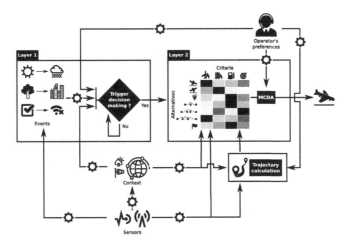

**Fig. 1.** Proposed multi-layer decision engine

**Layer 2** consists in determining which high-level action (e.g. *takeoff, continue, skip* one or more waypoints, *return to base, loiter, land, ...*) is the best answer to the risks generated by the event from the first layer. The evaluation of these actions is supplied by the context [1] but also by the trajectories that the drone will take. The context, which is taken into account within the second layer, is a set of elements that describes the UAV's environment. It can include information about the mission, its objective (e.g. to monitor a target, protect a convoy, ...) or its current state. Other information related to the drone are also included in this context which are given by the UAV's onboard sensors. The sensor outputs can be used directly (e.g. GPS coordinates, altitude) or they can be processed before, while other information regarding its surroundings can also be used (flight zone map, invisibility zone, weather conditions). A trajectory calculation module is also included, and can, for example, be implemented through the work of [16], where the computation is based on the operator's preferences.

In our case, we have retained four such consequences, or criteria, that a high-level action might have (but they could be more diverse):

- energy consumption [10], corresponding to the amount of energy left to complete the mission after executing the selected action,
- risk to the drone [20], i.e. the risk associated with flying over different areas such as forests, sea, military zones,
- risk to the environment, such as people, buildings, in the case of a crash,
- mission progress, which is a weighted percentage of the achieved sub-objectives.

Each operator may view these elements, i.e. the consequences of the possible decision actions, differently. As a result, a model of the operator's preferences has to be constructed prior to the mission. We propose to rank the different decision actions with respect to their evaluations on the multiple criteria and the preferences of the operator by integrating a multi-criteria decision model.

# 4  The Multi-Criteria Decision Aiding Model

As already mentioned, the DM is the operator, whose preferences must be modeled before including them in the decision engine of the autonomous drone. To guarantee a certain level of trust in the UAV's decision making process, the preference model and its consequences should be presented to the operator in order to be validated beforehand. It is therefore of high importance that this model is easy to explain to a non-expert of MCDA and that the decision recommendations (the recommended UAV actions which will ultimately influence the UAV's behavior) can be easily justified. We chose to implement a method called SRMP (Simple Ranking Method using Reference Profiles). This choice is motivated by the following 3 reasons: both qualitative and quantitative criteria can be easily integrated; the output is a pre-order of the alternatives, as the UAV may have to choose the second, or third-best alternative, in case he's not able to execute the best one; the output can be explained through a series of rules to the operator. The last point is of particular interest due to the critical nature of the decisions that an autonomous drone must make during operation.

## 4.1  Simple Ranking Method Using Reference Profiles (SRMP)

SRMP has originally been proposed by [18] and theoretically studied and characterized axiomatically by [5]. Reference points are used in the comparison of two alternatives: $a$ is considered as strictly preferred to $b$ if and only if the outranking relation between $a$ and the reference point is "stronger" than the outranking relation between $b$ and the reference point. Let us now show how this is implemented more formally.

We denote with $\mathcal{A}$ a set of $n$ alternatives and with $M = \{1, \ldots, m\}$ the indexes of $m$ criteria. The evaluation of an alternative $a \in \mathcal{A}$ on criterion $j \in M$ is denoted with $a_j$.

The SRMP method is defined by several preference parameters which need to be identified beforehand. These parameters are:

- the reference profiles: $\mathcal{P} = \{p^h, h = 1, \ldots, k\}$ where $p^h = \{p_1^h, \ldots, p_j^h, \ldots, p_m^h\}$ corresponds to the evaluations of $p^h$ on all criteria and $p_j^h \succsim_j p_j^l, \forall h, l \in \{1, \ldots, k\}, h > l$, and $\succsim_j$ representing the preferential pre-order on the values of criterion $j$;
- the lexicographic order of the profiles: $\sigma$, which corresponds to a permutation on $1, \ldots, k$;
- the criteria weights: $w_1, w_2, \ldots, w_m$, where $w_j \geq 0$ and $\sum_{j \in M} w_j = 1$

SRMP consists of a three-steps procedure as follows:

1. compute $C(a, p^h) = \{j \in M : a_j \succsim_j p_j^h\}$ with $a \in \mathcal{A}$, $h = 1, \ldots, k$, the set of criteria on which alternative $a$ is at least as good as profile $p^h$.
2. compare all pairs of alternatives $a, b \in \mathcal{A}$ to the reference profiles in order to define the following relations:

$$- a \succ_{p^h} b \Leftrightarrow \sum_{j \in C(a_i, p^h)} w_j > \sum_{j \in C(b, p^h)} w_j$$

$$- a \sim_{p^h} b \Leftrightarrow \sum_{j \in C(a_i, p^h)} w_j = \sum_{j \in C(b, p^h)} w_j$$

3. rank two alternatives $a, b \in \mathcal{A}$ by sequentially considering the relations $\succsim_{p^{\sigma(1)}}, \succsim_{p^{\sigma(2)}}, \ldots, \succsim_{p^{\sigma(k)}}$ (according to the lexicographic order $\sigma$):
   - $a$ is preferred to $b$ iff:

$$(a \succ_{p^{\sigma(1)}} b) \text{ or}$$

$$(a \sim_{p^{\sigma(1)}} b \text{ and } a \succ_{p^{\sigma(2)}} b) \text{ or}$$

$$\ldots$$

$$(a \sim_{p^{\sigma(1)}} b \text{ and } \ldots \text{ and } a \sim_{p^{\sigma(k-1)}} b \text{ and } a \succ_{p^{\sigma(k)}} b)$$

   - $a$ is indifferent to $b$ iff:    $a \sim_{p^{\sigma(1)}} b$ and $\ldots$ and $a \sim_{p^{\sigma(k)}} b$

### 4.2 Illustrative Example

Let us show on a small example how the UAV could use this SRMP model to make decisions. Imagine that the UAV has to select among 3 high level actions $x$, $y$ and $z$, like for example "land", "loiter" and "skip a waypoint", once the second layer of our decision engine has been triggered.

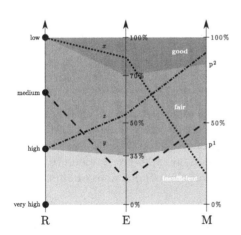

**Fig. 2.** SRMP example

**Table 1.** Evaluations of the alternatives and SRMP parameters.

|   | R | E | M |
|---|---|---|---|
| $x$ | Low | 80% | 20% |
| $z$ | Medium | 20% | 50% |
| $y$ | High | 55% | 90% |
| $p^1$ | High | 35% | 40% |
| $p^2$ | Low | 70% | 80% |
| $\sigma\{1, 2\}$ | | | |

With each of these actions, a trajectory is associated, which has been calculated beforehand by the trajectory calculation module. The three actions are evaluated on three criteria, the risk (R), the energy consumption (E) and the mission progress (M), and the result is presented in Table 1. The preference parameters of the SRMP model, which model the preferences of an operator, are also given in this table. They have been learned from a prior preference elicitation process such as the one we will present in Sect. 4.3.

The two reference profiles allow to define three intervals on the performances on each criterion: better than $p^2$; between $p^1$ and $p^2$; worse than $p^1$. This allows to identify intervals of performances as illustrated in Fig. 2, such that: "good" performances are above $p^2$, "intermediate" performances are between $p^1$ and $p^2$ on each criterion, "insufficient" performances are below $p^1$ on each criterion.

Let us now follow the steps presented earlier to rank the three alternatives $x$, $y$ and $z$. First we compute $C(a, p^h) = \{j \in M : a_j \geqslant p_j^h\}$, $\forall a \in \mathcal{A} = \{x, y, z\}$, $h \in \{1, 2\}$, $M = \{R, E, M\}$ and then compare each alternative to the others by using the profiles $p^h$, and finally rank the alternatives by considering the lexicographic order $\sigma = \{1, 2\}$

$p^1$:

$$\left. \begin{array}{l} \sum_{j \in C(x,p^1)} w_j = 1/3 + 1/3 + 0 = 2/3 \\ \sum_{j \in C(y,p^1)} w_j = 1/3 + 1/3 + 1/3 = 1 \\ \sum_{j \in C(z,p^1)} w_j = 1/3 + 0 + 1/3 = 2/3 \end{array} \right\} \Rightarrow \begin{array}{l} y \succ_{p^1} x \\ y \succ_{p^1} z \\ x \sim_{p^1} z \end{array}$$

$p^2$:

$$\left. \begin{array}{l} \sum_{j \in C(x,p^2)} w_j = 1/3 + 1/3 + 0 = 2/3 \\ \sum_{j \in C(z,p^2)} w_j = 0 + 0 + 0 = 0 \end{array} \right\} \Rightarrow x \succ_{p^2} z$$

The final ranking is thus $y \succ x \succ z$, hence $y$ is globally the best alternative, followed by $x$ and then $z$. This result can be explained to the operator in the following way: $y$ is better than all other alternative because it does not have any "insufficient" evaluations, while $x$ and $z$ have one "insufficient" evaluation on criterion M, respectively E; $x$ is better than $z$ because it has "good" evaluations on criteria R and E while $z$ does not have any "good" evaluations. The drone will thus implement decision $y$ and its corresponding trajectory.

### 4.3   Incremental Preference Elicitation for SRMP Models

Olteanu et al. propose in [17] to learn the preference parameters of SRMP models from a set of pairwise comparisons of alternatives given by a DM in one iteration. They thus formulate SRMP preference elicitation as a Mixed Integer linear Program (MIP), and show that to obtain an expressive preference model, this learning algorithm requires quite a few pairwise comparisons of alternatives as inputs.

In order to reduce the cognitive effort of the operator during the preference elicitation process, we propose an *incremental learning process* for SRMP models presented in Fig. 3, which should reduce the number of pairwise comparisons of alternatives that the operator has to evaluate. This process is performed before the mission, and its output (the preference parameters of the SRMP model) is then integrated into the second layer of our decision engine to configure the SRMP algorithm presented in Sect. 4.1. The input of the learning process is a database $\mathcal{D}$ of pairs of alternatives/decision actions (typically, two actions among which the autonomous UAV would have to choose during a mission). At each iteration a heuristic selects a pair of alternatives $(a, b)$ from $\mathcal{D}$ and the operator expresses his/her preferences by answering a *pair-wise comparison question*: do you strictly prefer alternative $a$ to alternative $b$, $b$ to $a$, or are you indifferent

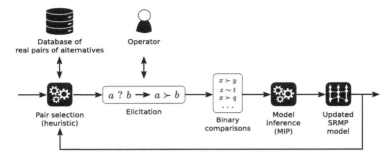

**Fig. 3.** Incremental learning process

between $a$ and $b$? Then they are added as a supplementary constraint in the MIP which infers the new SRMP model parameters. This procedure is repeated, as depicted by the continuous arrow until a "good enough" preference model is obtained. Note that in our approach we do not take into consideration any uncertainty in the operator's preferences.

The selection heuristic that we propose in this article, and which we name $\mathcal{H}_{mp}$, works as follows. The first iteration is a random selection of a pair of alternatives from $\mathcal{D}$. Then, at each iteration $i$, we use the preference model $M_{i-1}$ generated in the previous iteration to select the next pair of alternatives. The idea is to select a pair which, in the current model $M_{i-1}$ uses the highest possible number of profiles in its comparison (ideally a pair considered as indifferent by $M_{i-1}$). By confronting the operator with such a pair, we hope that his/her answer will generate a new constraint for the MIP which will reduce the size of the search space, and thus the possible values that the preference parameters may take.

**Table 2.** $\mathcal{H}_{mp}$ penalty function

| Penalty | Proposition |
|---|---|
| 1 | $(a \sim_{p^{\sigma(1)}} b$ and ... and $a \sim_{p^{\sigma(k)}} b)$ |
| 2 | $(a \sim_{p^{\sigma(1)}} b$ and ... and $a \sim_{p^{\sigma(k-1)}} b$ and $a \succ_{p^{\sigma(k)}} b)$ |
| ... | ... |
| k | $(a \sim_{p^{\sigma(1)}} b$ and $a \succ_{p^{\sigma(2)}} b)$ |
| k + 1 | $(a \succ_{p^{\sigma(1)}} b)$ |

We start by selecting a pair $(a, b)$ that is indifferent using $M_{i-1}$. This means that for $(a, b)$ all the $k$ profiles have been tested in the SRMP procedure, and model $M_{i-1}$ was not able to say whether $a$ is preferred to $b$ or $b$ is preferred to $a$. If there is no such indifferent pair, a pair that uses $k$ profiles in $M_{i-1}$ to be discriminated will be selected. If no such pair exists, we search for a pair using

$k-1$ profiles, and so on, until reaching the case of one profile. To model this more formally, let us associate an integer penalty score with the logical propositions of Table 2. For each pair $(a, b) \in \mathcal{D}$ we associate the penalty of the first proposition which is true in the sequence from Table 2. Finally, the $\mathcal{H}_{\mathrm{mp}}$ heuristic selects a pair $(a, b)$ from $\mathcal{D}$ that minimizes this penalty.

In order to validate empirically that the $\mathcal{H}_{\mathrm{mp}}$ selection heuristic allows to find a good preference model with a limited number of pairwise comparisons, we perform some experiments, and compare it to a random selection of the pairs from the database (we call this heuristic $\mathcal{H}_{\mathrm{rnd}}$).

These experiments follow the incremental learning process presented in Fig. 3 with an additional test phase to evaluate the quality of the obtained SRMP model. A database $\mathcal{D}$ of 100 pairs of alternatives is used as input for the proposed heuristic. $\mathcal{H}_{\mathrm{mp}}$ selects a pair of alternatives from $\mathcal{D}$ at each iteration. The operator of Fig. 3 is replaced for our experiments with a randomly generated SRMP model $M_{\mathrm{OP}}$. It is used to compare pairs of alternatives, which in turn generate a new constraints for model $M_i$.

To test the quality of a model, we generated a test database $D_{test}$ composed of 5000 alternatives. These alternatives are ranked both by the original SRMP model $M_{\mathrm{OP}}$ and the current one $M_i$. The quality of $M_i$ is then evaluated through *Kendall*'s rank correlation measure $\tau$ between these two rankings. $\tau$ measures the correlation of two rankings, and varies between 1 and $-1$. If both rankings are identical then $\tau = 1$, while if they are completely inverted then $\tau = -1$.

We execute this process for 100 different artificial databases $\mathcal{D}$, composed each of 100 pairs of alternatives, for different problem sizes ($m = 3, 5, 7$). We also fix the number of profiles to 2. This generates 3 problem configurations which we call (2P 3C), (2P 5C) and (2P 7C) (for $k$ profiles and $m$ criteria).

Each of the plots in Fig. 4 depicts the average value of the *Kendall tau* across the 100 different artificial databases as a function of the number of pairwise comparisons submitted to $M_{\mathrm{OP}}$. For example, for problems containing 3 criteria, after asking the $M_{\mathrm{OP}}$ to compare 40 pairs of alternatives, selected with the $\mathcal{H}_{\mathrm{mp}}$ heuristic, we can obtain on average a preference model which ranks the test data quite similarly to the way $M_{\mathrm{OP}}$ ($\tau \sim 85\%$).

What we can observe here is that the Kendall $\tau$ increases when adding new preferences of the operator (i.e. pairs of alternatives). The plots show that the $\mathcal{H}_{\mathrm{mp}}$ heuristic dominates the $\mathcal{H}_{\mathrm{rnd}}$ one, which is confirmed by *Kolmogorov-Smirnov* statistical tests allowing to compare two samples [8]. We also notice that for the first few iterations both curves ($\mathcal{H}_{\mathrm{mp}}$ and $\mathcal{H}_{\mathrm{rnd}}$) behave similarly (for the different problem sizes), which is due to the small number of learning pairs involved, and which tend to produce not very expressive SRMP models. Then, both curves separate clearly in favour of $\mathcal{H}_{\mathrm{rnd}}$. For the last few iterations, the curves become again similar, which is explained by the fact that the set of learning pairs is almost the same, independently of the selection heuristic ($\mathcal{D}$ is finite and fixed to 100 pairs of alternatives).

The standard deviations associated with the average values depicted in these figures are small (on average $\sim 0.075$ for 2P 3C for both $\mathcal{H}_{\mathrm{mp}}$ and $\mathcal{H}_{\mathrm{rnd}}$, $\sim 0.095$

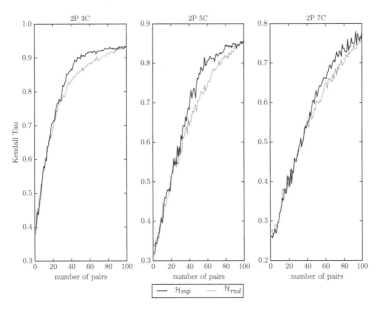

**Fig. 4.** Average Kendall tau for 2P 3C, 2P 5C and 2P 7C

for 2P 5C for both $\mathcal{H}_{mp}$ and $\mathcal{H}_{rnd}$, and $\sim$0.109 for 2P 7C for both $\mathcal{H}_{mp}$ and $\mathcal{H}_{rnd}$) and they also decrease with the addition of more learning pairs. They have not been included in these illustrations for these reasons.

This result can be used in a practical case and gives an answer to the research question which is how many learning pairs/iterations are needed to achieve a *"good"* SRMP model with an objective to reduce the cognitive effort of the operator. An SRMP model is considered as a *"good"* one by the operator if this model can reach a given *Kendall tau* value. Once this valus is given we use the average *Kendall tau* curves to find the number of pairs required to reach $\tau$.

For example, Fig. 5 depicts the average *Kendall tau* for *2P 3C* where the operator fixed $\tau = 0.9$ we can see that we need about 48 learning pairs by using the $\mathcal{H}_{mp}$ heuristic while about 72 for the $\mathcal{H}_{rnd}$ heuristic.

## 5    Experimental Validation of the Decision Model

In order to validate our proposal we developed a program which simulates the flight of a UAV containing the previously presented decision engine. The graphical user interface (GUI) of the simulator is presented in Fig. 6.

The modeled autonomous UAV is a Watchkeeper Unmanned Aircraft System from Thales [23], represented by a point which is submitted to physical constraints. The simulated UAV is able to navigate through a set of waypoints and execute different high-level actions (e.g. take off, loiter, ...). The simulator is also able to evaluate these actions on different criteria presented in Sect. 3.

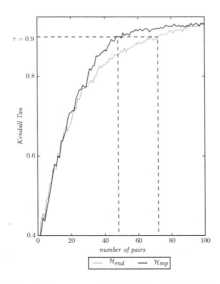

**Fig. 5.** Average Kendall tau for 2P 3C

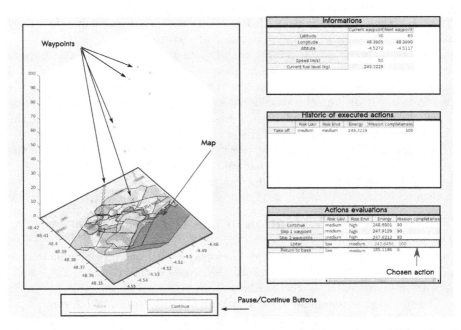

**Fig. 6.** Graphical user interface for the autonomous UAV simulator (Color figure online)

The GUI presented in Fig. 6 is composed of four parts. The left panel plots the details of the mission (different waypoints (as red crosses) and mission map). It also shows two maps of the risk associated with the UAV and the environment respectively, which allow the evaluation of the risk of a trajectory using a weighted average of the risk of the different zones overflown by the drone. The top right box provides information about the current waypoint and the next ones, as well as information on the current speed and the amount of energy left. The middle right box presents a history of the executed actions. Finally, the bottom right panel shows the evaluations of all the possible high-level actions for the current waypoint. The SRMP model is executed in the background in order to decide which action will be chosen next (highlighted in green).

To illustrate our work, we provide here an example, where the UAV has to accomplish a mission consisting of flying through a set of nine waypoints and taking photos at each of them. We suppose that for waypoints 1, 5 and 6, these photos are missed, which requires the UAV to loiter for a second shot in order to complete the mission at 100%. We execute the mission according to two different operator profiles, represented by two different sets of preference parameters.

**Table 3.** Preference parameters for the two operators.

|  | Operator 1 | | | | Operator 2 | | | |
|---|---|---|---|---|---|---|---|---|
|  | $R_{UAV}$ | $R_{Env}$ | E | M | $R_{UAV}$ | $R_{Env}$ | E | M |
| $p^1$ | High | v.high | 30% | 30% | High | v.high | 30% | 30% |
| $p^2$ | Low | Medium | 60% | 99% | Low | Medium | 60% | 70% |
| $w$ | 0.1 | 0.1 | 0.1 | 0.7 | 0.25 | 0.25 | 0.25 | 0.25 |
| $\sigma$ | $\{2,1\}$ | | | | $\{2,1\}$ | | | |

The first operator is mainly focusing on completing the mission, placing as secondary objectives the risk and the fuel consumption. The incremental preferences elicitation phase (Sect. 4.3) leads to the SRMP parameters presented in the left half of Table 3. The second second operator gives a more uniform importance to the mission completeness, risks and fuel consumption objectives. Those preference parameters are summarized in the right half of Table 3, and have again been obtained using our incremental elicitation process.

As expected the UAV configured with first operator's preferences accomplishes the mission with success, as shown in Fig. 7 on the left, by flying through all the waypoints and loitering at waypoints 1, 5 and 6 in order to take another round of photos, without taking into account the risk linked to the underlying zones. The right side of Fig. 7 shows the execution of the mission with respect to the preferences of the second operator. We can observe that the UAV, even if only one photo was taken at waypoints 5 and 6, did not loiter (because it considered it too risky to fly again over the same zone), and even skipped waypoint 6 (for the same reason).

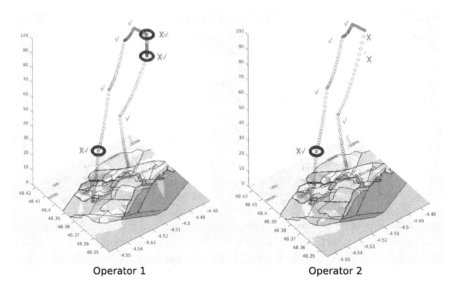

**Fig. 7.** Mission simulation with the preferences of operators 1 and 2

## 6    Conclusion and Perspectives

In this work, we propose a new approach for integrating an operator's perspective within the decision engine of autonomous UAVs, through a multi-layer decision engine, a traceable MCDA technique and an incremental process which minimizes the cognitive effort of the operator during the preference elicitation. Dividing the decision process of the autonomous UAV into several layers allows us to integrate the perspective of the operator in different elements of the autonomous decision making process, and thus provides an autonomy of the UAV guided by the preferences of a human operator. Depending on the characteristics of the decision problem (as for example the number of considered criteria), the resolution of the MIP which is used iteratively in the elicitation process can take some time. This could limit its use in practice in an incremental elicitation process, which motivates us to study in a next step approximate algorithms (meta-heuristics) for the determination of the parameters of the SRMP model.

## References

1. Ajami, A., Balmat, J.F., Gauthier, J.P., Maillot, T.: Path planning and ground control station simulator for UAV. In: IEEE Aerospace Conference, pp. 1–13 (2013)
2. Arantes, M.S., Arantes, J.S., Toledo, C.F.M., Williams, B.C.: A hybrid multi-population genetic algorithm for UAV path planning. In: Proceedings of the Genetic and Evolutionary Computation Conference 2016, GECCO 2016, pp. 853–860. ACM, New York (2016)

3. Benabbou, N., Perny, P., Viappiani, P.: Incremental elicitation of Choquet capacities for multicriteria choice, ranking and sorting problems. Artif. Intell. **246**, 152–180 (2017)
4. Blackmore, L., Ono, M., Williams, B.C.: Chance-constrained optimal path planning with obstacles. IEEE Trans. Robot. **27**(6), 1080–1094 (2011)
5. Bouyssou, D., Marchant, T.: Multiattribute preference models with reference points. Eur. J. Oper. Res. **229**(2), 470–481 (2013)
6. Ciomek, K., Kadziński, M., Tervonen, T.: Heuristics for selecting pair-wise elicitation questions in multiple criteria choice problems. Eur. J. Oper. Res. **262**(2), 693–707 (2017)
7. Delmerico, J., Mueggler, E., Nitsch, J., Scaramuzza, D.: Active autonomous aerial exploration for ground robot path planning. IEEE Robot. Autom. Lett. **2**(2), 664–671 (2017)
8. Dodge, Y.: Kolmogorov–Smirnov Test, pp. 283–287. Springer, New York (2008)
9. Durbach, I.: The use of the SMAA acceptability index in descriptive decision analysis. Eur. J. Oper. Res. **196**(3), 1229–1237 (2009)
10. Franco, C.D., Buttazzo, G.: Energy-aware coverage path planning of UAVs. In: 2015 IEEE International Conference on Autonomous Robot Systems and Competitions, pp. 111–117, April 2015
11. Holloway, H., White III, C.C.: Question selection for multi-attribute decision-aiding. Eur. J. Oper. Res. **148**(3), 525–533 (2003)
12. Jacquet-Lagreze, E., Siskos, J.: Assessing a set of additive utility functions for multicriteria decision-making, the UTA method. Eur. J. Oper. Res. **10**(2), 151–164 (1982)
13. Kabamba, P., Meerkov, S., Zeitz, F.: Optimal path planning for unmanned combat aerial vehicles to defeat radar tracking. J. Guid. Control Dyn. **29**(2), 279–288 (2006)
14. Kim, D., Chen, T.: Deep neural network for real-time autonomous indoor navigation. CoRR abs/1511.04668 (2015). http://arxiv.org/abs/1511.04668
15. Lahdelma, R., Hokkanen, J., Salminen, P.: SMAA - stochastic multiobjective acceptability analysis. Eur. J. Oper. Res. **106**(1), 137–143 (1998)
16. Narayan, P., Meyer, P., Campbell, D.: Embedding human expert cognition into autonomous UAS trajectory planning. IEEE Trans. Cybern. **43**(2), 530–543 (2013)
17. Olteanu, A.L., Belahcène, K., Mousseau, V., Ouerdane, W., Rolland, A., Zheng, J.: Preference elicitation for a ranking method based on multiple reference profiles, August 2018. Working paper, https://hal.archives-ouvertes.fr/hal-01862334
18. Rolland, A.: Reference-based preferences aggregation procedures in multi-criteria decision making. Eur. J. Oper. Res. **225**(3), 479–486 (2013)
19. Roy, B.: Multicriteria Methodology for Decision Aiding. Kluwer Academic, Dordrecht (1996)
20. Ruz, J.J., Arevalo, O., de la Cruz, J.M., Pajares, G.: Using MILP for UAVs trajectory optimization under radar detection risk. In: 2006 IEEE Conference on Emerging Technologies and Factory Automation, pp. 957–960, September 2006
21. Sadeghi, F., Levine, S.: Real single-image flight without a single real image. CoRR abs/1611.04201 (2016). http://arxiv.org/abs/1611.04201
22. Taisho, T., Enfu, L., Kanji, T., Naotoshi, S.: Mining visual experience for fast cross-view UAV localization. In: 2015 IEEE/SICE International Symposium on System Integration (SII), pp. 375–380, December 2015

23. Thales Group: Watchkeeper (2010). https://bit.ly/2IY6LyJ
24. The, A., Mousseau, V.: Using assignment examples to infer category limits for the ELECTRE TRI method. J. Multi-Criteria Decis. Anal. **11**(1), 29–43 (2002)
25. Tipaldi, M., Glielmo, L.: A survey on model-based mission planning and execution for autonomous spacecraft. IEEE Syst. J. **PP**(99), 1–13 (2017)

# Refugee Allocation in the Setting of Hedonic Games

Benno Kuckuck[1], Jörg Rothe[2(✉)], and Anke Weißenfeld[2,3]

[1] Mathematisches Institut, Heinrich-Heine-Universität Düsseldorf, Düsseldorf, Germany
benno.kuckuck@hhu.de
[2] Institut für Informatik, Heinrich-Heine-Universität Düsseldorf, Düsseldorf, Germany
[3] Ministry for Children, Family, Refugees and Integration of the State of North Rhine-Westphalia, Düsseldorf, Germany
{rothe,anke.weissenfeld}@hhu.de

**Abstract.** In recent work, Aziz et al. [4] consider refugee allocation as a matching problem, akin to the well-known hospitals-residents problem. They consider a wide range of stability conditions. Hedonic games are a well-studied class of coalition formation games, that encompass the classical matching problems. We propose a transformation of the Refugee Allocation Problem as formulated by Aziz et al. [4] into the setting of hedonic games, parametrized by a set extension rule. We show that different set extension rules lead to different stability concepts, derived from the central concept of core stability in hedonic games, mirroring some of the stability concepts proposed by Aziz et al. [4].

**Keywords:** Matching problem · Hedonic game · Refugee allocation

## 1 Introduction

Matching problems have long been of central importance in economics. Gale and Shapley [14] were the first to study the stable matching (or stable marriage) problem: There are $n$ male and $n$ female players, the males rank the females according to their preferences and the females rank the males according to their preferences, the goal being to form man-woman couples in a "stable" manner. Since their seminal paper, the concept of stable matchings has been famously used in the U.S. National Residents Matching Program to place medical school students into hospital residency training programs as well as to model aspects of labor markets more broadly (see the work of Roth and Sotomayor [25], Hatfield and Milgrom [16], and Roth [24] for introductions to the topic). As a consequence, many extensions and generalizations of the classical problems have been studied. For example, the stable roommate problem [18] does not have a bipartite structure as the stable matching problem does: There are $2n$ unisex players, each ranking the other players according to their preferences, and all possible pairs are feasible. Another generalization proposed by Alkan [1] considers three "genders"

This work was supported in part by DFG grant RO 1202/14-2.

S. Pekeč and K. B. Venable (Eds.): ADT 2019, LNAI 11834, pp. 65–80, 2019.
https://doi.org/10.1007/978-3-030-31489-7_5

(say, men, women, and dogs) who each rank the other "genders," and now we are looking for stable threesomes. Aziz et al. [4] propose yet another such extension involving heterogeneous, multi-dimensional demands and capacities, which they frame as a
problem of assigning refugee families to localities (based on earlier work by Biró and
McDermid [9] and Delacrétaz et al. [11]). Their approach will be central to this paper.

Coalition formation games have likewise long played an important role in economics and game theory. More recently, hedonic games have provided a popular basic
framework for studying questions of computational complexity surrounding stability
(see the book chapters by Aziz and Savani [5] and Elkind and Rothe [13] and the survey by Woeginger [26] for more background and pointers to the literature).

In particular, Woeginger [26] and Aziz [2] discuss how classical matching problems
such as the stable matching problem, the stable roommate problem, and the stable threesome problem mentioned above can be reformulated as hedonic games. This approach
has the advantage that it makes concepts and results from the theory of hedonic games –
such as various types of stability – applicable to these matching problems.

In this paper, we translate the refugee allocation problem of Aziz et al. [4] to the
framework of hedonic games. We compare their concepts of stability to well-known
stability notions from hedonic games, showing how different ways of translating preferences over individual agents to preferences over sets of agents to some extent mirror
the difference in stability concepts considered in Aziz et al. [4].

This paper is organized as follows: In Sect. 2, we provide the basic notions and
definitions from matching theory, focusing on stability notions. In particular, we define
the refugee allocation problem of Aziz et al. [4] in detail and consider various properties
of allocations. We present the refugee allocation problem in the fairly general context
of what we call matching problems with feasibility constraints. Most of the definitions
from Aziz et al. [4] readily generalize to this context and are potentially interesting not
just in the narrower model of refugee allocation problems.

In Sect. 3, we give the needed notions from the theory of hedonic games and then
formulate the refugee allocation problem of Aziz et al. [4] as a hedonic game. To this
end, we also need to consider set extensions so as to lift preferences over a set to preferences over all its subsets. Next, we present the technical results of this paper, regarding
stability in refugee allocation and the corresponding hedonic games.

Finally, we conclude in Sect. 4 with a brief outlook on future work.

## 2   Basics

In this section, we describe the refugee allocation problems introduced by Aziz et al. [4]
and then define various properties of refugee allocation problems including their stability. We first give a more general definition of a matching problem with arbitrary feasibility constraints, since virtually all of the notions used in Aziz et al. [4] readily generalize to this context. See, e.g., [15, 19] for similar (but not directly comparable) general matching problems with constraints, or [16, 22, 23, 25] for an overview of matching
problems with complex preferences.

## 2.1   Matching Problems with Feasibility Constraints

A matching problem with feasibility constraints (FeaCoMP) is a tuple $\mathcal{M} = (F, L, \succsim, \Phi)$ where $F$ is a finite set of *refugee families*, $L$ is a finite set of *localities*, $\Phi = (\Phi_l)_{l \in L}$ is a tuple of *feasibility sets* $\Phi_l \subseteq 2^F$, closed under taking subsets, for each $l \in L$ specifying which sets of families may be assigned to locality $l$, and, finally, $\succsim = ((\succsim_f)_{f \in F}, (\succsim_l)_{l \in L})$ is a *preference profile*, consisting of weak orders $\succsim_f$ over $L \cup \{\emptyset\}$ for each $f \in F$ and weak orders $\succsim_l$ over $F \cup \{\emptyset\}$ for each $l \in L$. In all of these orderings, $\emptyset$ (which we assume to be an element of neither $F$ nor $L$) is to be interpreted as a dummy element, separating the acceptable choices for each family or locality, which are ranked above $\emptyset$, from the unacceptable ones, which are ranked below $\emptyset$. If all of the $\succsim_p$ with $p \in F \cup L$ are in fact total orders, we say that $\mathcal{M}$ is a FeaCoMP *without indifferences*.

We will denote for a weak order (or, more generally, a quasi-order) $\succsim$ on a set $X$ the corresponding indifference relation by $\sim$ (i.e., for $x, y \in X$ it holds that $x \sim y$ if and only if $x \succsim y$ and $y \succsim x$) and the corresponding strict order by $\succ$ (i.e., for $x, y \in X$ it holds that $x \succ y$ if and only if $x \succsim y$ and not $y \succsim x$). A map $\pi \colon F \to L \cup \{\emptyset\}$ is called an allocation for $\mathcal{M}$ as above, where $\pi(f) = \emptyset$ is used to indicate that $f$ is not allocated to any locality. We will usually abbreviate the set $\pi^{-1}(\{l\})$ of families allocated to a locality $l \in L$ by $\pi^{-1}(l)$. Formally, such a map is a subset $\pi \subseteq F \times (L \cup \{\emptyset\})$, and so we will occasionally use the notation $(f, l) \in \pi$ to denote $f \in F$ being a family with $l = \pi(f)$. An allocation $\pi$ for $\mathcal{M}$ is called *feasible* if $\pi^{-1}(l) \in \Phi_l$ for all $l \in L$. If the FeaCoMP is clear from the context we simply say that $Y \subseteq F$ is feasible for $l \in L$ if $Y \in \Phi_l$.

## 2.2   Refugee Allocation Problems

When considering decision problems about FeaCoMPs, there is of course the problem (shared with hedonic games, which we consider later), that representing the feasibility sets might require space exponential in the number of families. Hence, for computational questions, we will usually have to find efficient representations for (particular families of) FeaCoMPs first. Refugee allocation problems, as defined by Aziz et al. [4] can be regarded as one such representation. A *refugee allocation problem* (RAP, for short), in the sense of Aziz et al. [4], is represented by a tuple $\mathcal{R} = (F, L, \succsim, S, d, c)$, where $F$, $L$, and $\succsim$ are as in the definition of FeaCoMPs above, $S$ is a finite set of *services*, $d = (d_f^s)_{(f,s) \in F \times S}$ specifies the demand $d_f^s \in \mathbb{N}_0$ each family $f \in F$ has for each service $s \in S$, $c = (c_l^s)_{(l,s) \in L \times S}$ specifies the *capacity* each locality $l \in L$ has for service $s \in S$. Given a RAP $\mathcal{R}$ as above, we denote by $d_f = (d_f^s)_{s \in S}$ the vector of demands for a family $f \in F$ and by $c_l = (c_l^s)_{s \in S}$ the vector of service capacities for a locality $l \in L$. The matching problem with feasibility constraints represented by $\mathcal{R} = (F, L, \succsim, S, d, c)$ is $\mathcal{M}(\mathcal{R}) = (F, L, \succsim, \mathcal{F})$ where we define the feasibility sets $\mathcal{F}_l$ for each $l \in L$ by letting $Y \in \mathcal{F}_l$ for $Y \subseteq 2^F$ if and only if $\sum_{f \in Y} d_f^s \leq c_l^s$ for all $s \in S$ (or $\sum_{f \in Y} d_f \leq c_l$, for short).

The hospitals-residents problem appears as the special case where $S$ is a singleton set (the only service that hospitals supply to residents being a position), and all demand vectors equal to $(1)$ (each resident requiring exactly one position).

## 2.3  Properties of Allocations

A common property studied in matching problems is individual rationality, demanding that a matching only contain pairs of agents that consider each other acceptable.

**Definition 1.** *An allocation $\pi$ for a FeaCoMP $\mathcal{M} = (F, L, \succsim, \Phi)$ is individually rational if it holds for all $(f, l) \in \pi$ that $l \succsim_f 0$ and $f \succsim_l 0$.*

Another important quality for a matching to be considered desirable is that no locality $l$ has unused service capacities while there are still families that could feasibly join $l$ and would prefer to do so. Aziz et al. [4] call this property *non-wastefulness*:

**Definition 2.** *A feasible allocation $\pi$ for the FeaCoMP $\mathcal{M} = (F, L, \succsim, \Phi)$ is wasteful if there exists a pair $(f, l) \in F \times L$ such that $l \succ_f \pi(f)$ as well as $f \succ_l 0$ and $\pi^{-1}(l) \cup \{f\}$ is feasible for $l$. Otherwise, $\pi$ is non-wasteful.*

Finding allocations that are merely individually rational and non-wasteful is easy (considering the families in turn, put each family in its favorite locality among those which find it acceptable and still have room). Aziz et al. [4] define a large number of stability concepts, all specializing to the classic stability concept for hospitals-residents problems in the single-service-equal-demands case. In general, an allocation is called stable if there is no pair of a family and a location which are currently not matched to each other, but would be able and inclined to deviate—a so-called blocking pair. The differences in stability concepts stem from the question of how to define "able and inclined." For example, would a locality consider (and be allowed to) remove several currently allocated families in favor of a preferred family, or just one? Would it do so even if it meant a larger strain on its service capacities? In the following we will present two of the stability concepts Aziz et al. [4] defined: *stability* and *weak stability*.

For weak stability, we assume that a pair $(f, l) \in F \times L$ blocks an allocation $\pi$ if $f$ prefers $l$ to its assigned locality under $\pi$ and the locality $l$ favors $f$ over at least one of its currently assigned families, $f'$, and could accommodate $f$ once it removes $f'$.

**Definition 3.** *Let $\pi$ be a feasible allocation for a FeaCoMP $\mathcal{M} = (F, L, \succsim, \Phi)$. We call $(f, l) \in F \times L$ a strongly blocking pair for $\pi$ if the following holds: (i) $l \succ_f \pi(f)$ and (ii) there is an $f' \in \pi^{-1}(l)$ such that $f \succ_l f'$ and $(\pi^{-1}(l) \setminus \{f'\}) \cup \{f\}$ is feasible for $l$. We call $\pi$ weakly stable if $\pi$ is non-wasteful and individually rational and there is no strongly blocking pair for $\pi$.*

For the stronger notion of stability, we say a pair $(f, l) \in F \times L$ blocks an allocation $\pi$ if $f$ prefers $l$ to its assigned locality under $\pi$ and the locality $l$ could accommodate $f$ once it removes all of its currently assigned families, which it likes less than $f$.

**Definition 4.** *Let $\pi$ be a feasible allocation for a FeaCoMP $\mathcal{M} = (F, L, \succsim, \Phi)$. We call $(f, l) \in F \times L$ a blocking pair for $\pi$ if the following holds: (i) $l \succ_f \pi(f)$ and (ii) there is a set $Y \subseteq \pi^{-1}(l)$ such that $f \succ_l f'$ for all $f' \in Y$ and $(\pi^{-1}(l) \setminus Y) \cup \{f\}$ is feasible for $l$. We call $\pi$ stable if $\pi$ is non-wasteful and individually rational and there is no blocking pair for $\pi$.*

Aziz et al. [4] also consider further stability notions that use an external *master list*. These stability notions assume that a locality can only swap a previously assigned family $f'$ for a preferred family $f$ if $f$ is higher on the master list than $f'$.

**Definition 5.** *Let* $\mathcal{M} = (F, L, \succsim, \Phi)$ *be a FeaCoMP, let* $\succsim_{\mathfrak{M}}$ *be a weak order on F, and let* $\pi$ *be a feasible allocation for* $\mathcal{M}$. *We call* $(f, l) \in F \times L$ *a strong blocking pair respecting* $\succsim_{\mathfrak{M}}$ *if (i)* $l \succ_f \pi(f)$, *and (ii) there is an* $f' \in \pi^{-1}(l)$ *such that* $f \succ_l f'$, $f \succsim_{\mathfrak{M}} f'$, *and* $(\pi^{-1}(l) \setminus \{f'\}) \cup \{f\}$ *is feasible for l; and we call* $(f, l) \in F \times L$ *a blocking pair respecting* $\succsim_{\mathfrak{M}}$ *if (i)* $l \succ_f \pi(f)$, *and (ii) there is a set* $Y \subseteq \pi^{-1}(l)$ *such that* $f \succ_l f'$ *and* $f \succsim_{\mathfrak{M}} f'$ *for all* $f' \in Y$ *and* $(\pi^{-1}(l) \setminus Y) \cup \{f\}$ *is feasible for l. The allocation* $\pi$ *is stable (respectively,* weakly stable*) with respect to master list* $\succsim_{\mathfrak{M}}$ *if* $\pi$ *is individually rational, non-wasteful, and there does not exist a blocking pair (respectively, a strong blocking pair) with respect to* $\succsim_{\mathfrak{M}}$.

The notion of weak stability with respect to a master list is interesting, because Aziz et al. [4] show that an allocation satisfying this condition exists and can be found in polynomial time for any RAP and any master list $\succsim_{\mathfrak{M}}$ with the restriction that $f \sim_{\mathfrak{M}} f'$ only holds if the demand vectors $d_f$ and $d_{f'}$ are identical.

In the case of FeaCoMPs we can replace this restriction as follows:

**Proposition 1.** *Let* $\mathcal{M} = (F, L, \succsim, \Phi)$ *be a FeaCoMP and let* $\succsim_{\mathfrak{M}}$ *be a weak order on F. Assume that the following exchangeability property holds: If* $l \in L$ *is a locality,* $Y \in \Phi_l$ *is feasible for l, and* $f' \in Y$, $f \in F$ *are families with* $f \sim_{\mathfrak{M}} f'$, *then* $(Y \setminus \{f'\}) \cup \{f\}$ *is likewise feasible for l. Then there exists a feasible allocation for* $\mathcal{M}$ *which is weakly stable with respect to* $\succsim_{\mathfrak{M}}$.

Such an allocation can be found by using the HFPDA algorithm proposed by Aziz et al. [4] and the proof of stability is virtually identical to the case for RAPs given in that paper, so we do not repeat it here.

# 3 A Fresh Look at the RAP

Based on the work of Delacrétaz et al. [11], Aziz et al. [4] considered the refugee allocation problem as a generalization of the *hospitals-residents problem* and established positive results by making suitable assumptions. Our goal is to formulate the refugee allocation problem (and matching problems with feasibility constraints, more generally) in another well-studied framework, namely in the context of hedonic games, which provides potential for further extensions and results. The fact that matching problems may be considered as hedonic (or NTU-)games is well-known and this connection has often been employed successfully (see, e.g., [7, 8, 12, 21, 26]). For more background on hedonic games, we refer to the book chapters by Aziz and Savani [5] and Elkind and Rothe [13] and to the survey by Woeginger [26].

## 3.1 Hedonic Games

A hedonic game is a pair $\mathcal{H} = (P, \succsim)$ consisting of a finite set of players (or agents) $P = \{p_1, \ldots, p_n\}$ and a *preference profile* $\succsim = (\succsim_p)_{p \in P}$, where $\succsim_p$ represents the preference list of agent $p$. A preference list $\succsim_p$ is a quasi-ordering[1] over the set $\{k \in 2^P \mid p \in k\}$ of all possible coalitions containing $p$.

The outcome or goal of such a game is to form a *coalition structure* or *partition* $\Gamma = \{k_1, \ldots, k_m\}$ of $P$ (i.e., $P = k_1 \cup \cdots \cup k_m$ and for all $i, j \in \{1, \ldots, m\}$, $i \neq j$, we have $k_i \cap k_j = \emptyset$). Denote by $\Gamma(p)$ the unique coalition containing $p$ in $\Gamma$.

We are interested in forming stable coalition structures. Among the many stability concepts known for hedonic games, we are in particular interested in the notion of *core stability*.

**Definition 6.** *Let $\mathcal{H}$ be a hedonic game and $\Gamma$ a coalition structure regarding $\mathcal{H}$. Then $\hat{k} \subseteq P$ is a* blocking coalition *if* $(\forall p \in \hat{k}: \hat{k} \succsim_p \Gamma(p)) \wedge (\exists \hat{p} \in \hat{k}: \hat{k} \succ_{\hat{p}} \Gamma(\hat{p}))$. *$\Gamma$ is* core stable *if there is no blocking coalition.*

For our purposes, we need to extend the common notion of hedonic game by the attribute of *feasible coalitions*, see, e.g., the work of Igarashi and Elkind [17]: A *hedonic game with feasible coalitions* is a triple $\mathcal{H}^{\mathcal{F}} = (P, \mathcal{F}, \succsim)$, still with players $P$ but in addition with a set $\mathcal{F} \subseteq 2^P$ of feasible coalitions, containing all the singleton coalitions $\{p\}$ for $p \in P$. The preference profile $\succsim = (\succsim_p)_{p \in P}$ now consists of quasi-orderings $\succsim_p$ only over $\{k \in \mathcal{F} \mid p \in k\}$. Likewise, a coalition structure $\Gamma$ of $P$ must now consist of feasible coalitions only, i.e., $k \in \mathcal{F}$ for each $k \in \Gamma$. For the notion of core stability, this means that a new coalition $\hat{k} \notin \Gamma$ can put a coalition structure $\Gamma$ at risk only if $\hat{k}$ itself is feasible. Core stability implies individual stability, which implies individual rationality.

**Definition 7.** *Let $\mathcal{H}^{\mathcal{F}}$ be a hedonic game with feasible coalitions and $\Gamma$ a coalition structure regarding $\mathcal{H}^{\mathcal{F}}$. Then $\Gamma$ is called* individually rational *if there does not exist a player $p \in P$ such that if $\{p\} \succ_p \Gamma(p)$ for all $p \in P$. Furthermore, $\Gamma$ is called* individually stable *if there do not exist a player $p \in P$ and a coalition $k \in \Gamma \cup \{\emptyset\}$ such that (i) $k \cup \{p\} \in \mathcal{F}$, (ii) $k \cup \{p\} \succ_p \Gamma(p)$, and (iii) for all $\hat{p} \in k$ it holds that $k \cup \{p\} \succsim_{\hat{p}} k$. Finally, $\hat{k} \in \mathcal{F}$ is a* blocking coalition *for $\Gamma$ if $(\forall p \in \hat{k}: \hat{k} \succsim_p \Gamma(p)) \wedge (\exists \hat{p} \in \hat{k}: \hat{k} \succ_{\hat{p}} \Gamma(\hat{p}))$, and $\Gamma$ is* core stable *if there is no blocking coalition in $\mathcal{F}$.*

## 3.2    RAP as a Hedonic Game

Now we can represent a RAP or, more generally, a FeaCoMP $\mathcal{M} = (F, L, \succsim, \Phi)$ as a hedonic game: The players are the refugee families and the localities. The coalitions will each consist either of a locality and the set of refugee families assigned to it or of a single (unassigned) family. Accordingly, only such coalitions are defined as feasible. Next, we have to define the preference profile for each locality and each refugee family. The original preference lists $\succsim$ are weak orderings over single players; we, however, will need (quasi-)orderings over coalitions. We approach this issue differently for the refugee families and for the localities. Since a refugee family in our model only has preferences over localities (not over other families), we assume that it is indifferent among all coalitions that feature the same locality. For the *localities* it is a bit more

---

[1] In the literature it is common to assume that the preferences are at least weak orders; however, we will have to consider also orders that are not complete.

difficult to define a preference over its feasible coalitions. The feasible coalitions containing $l$ consist of $l$ along with a set of refugee families. So, from a preference list of $l$ over $F$ we have to create a preference list over the *set of feasible subsets of $F$*. This is commonly referred to as a *set extension* of the ordering $\succsim_l$. We take a closer look at concrete set extension rules in the subsequent section. Here, we simply consider a set extension rule $\Xi$ to be a function mapping any quasi-ordering $\succsim$ on a set $X$ to a quasi-ordering $\succsim^{\Xi}$ on $2^X$.

**Definition 8.** *Let* $\mathcal{M} = (F, L, \succsim, \Phi)$ *be a FeaCoMP and $\Xi$ be a set extension rule. Define the associated hedonic game $\mathcal{H}(\mathcal{M}, \Xi) = (P, \mathcal{F}, \succsim^{\star})$ as follows:*

(a) $P = F \cup L$.

(b) *A coalition $k \subseteq P$ is in $\mathcal{F}$ if it is either of the form $\{f\}$ with $f \in F$ or it contains exactly one locality $l \in L$ and a subset of families $F' \subseteq F$ (which can also be empty) such that $F'$ is feasible for $l$: $\mathcal{F} = \{\{f\} \mid f \in F\} \cup \{\{l\} \cup F' \mid F' \text{ is feasible for } l\}$.*

(c) *For $f \in F$, a weak order $\succsim^{\star}_f$ is defined as follows: For any two feasible coalitions of the form $k = \{l, f\} \cup F'$ and $\hat{k} = \{\hat{l}, f\} \cup F''$ with $l, \hat{l} \in L$ and $F', F'' \subseteq F \setminus \{f\}$, we set $k \succsim^{\star}_f \hat{k} \iff l \succsim_f \hat{l}$. Furthermore, $k \succsim^{\star}_f \{f\} \iff k \succsim_f \emptyset$, and $\{f\} \succsim^{\star}_f k \iff \emptyset \succsim_f k$.*

(d) *For $l \in L$, we define a quasi-order $\succsim^{\star}_l$ as follows: For any two feasible coalitions of the form $k = \{l\} \cup F'$ and $\hat{k} = \{l\} \cup F''$ with $F', F'' \subseteq F$ we set $k \succsim^{\star}_l \hat{k} \iff (\exists f \in F'' : \emptyset \succ_l f) \vee ((\forall f \in F' : f \succsim_l \emptyset) \wedge (F' \succsim^{\Xi}_l F''))$. In words: The locality $l$ is indifferent between all coalitions that contain at least one family which is inacceptable to $l$, and strictly prefers all other coalitions to these; the coalitions containing only acceptable families are ordered according to the set extension rule $\Xi$.*

If $\mathcal{R}$ is a RAP, we write $\mathcal{H}(\mathcal{R}, \Xi)$ instead of $\mathcal{H}(\mathcal{M}(\mathcal{R}), \Xi)$ to denote the hedonic game associated to $\mathcal{M} = \mathcal{M}(\mathcal{R})$. If $\pi$ is a feasible allocation for the FeaCoMP $\mathcal{M}$, then $\Gamma = \{\{l\} \cup \pi^{-1}(l) \mid l \in L\} \cup \{\{f\} \mid f \in F, \pi(f) = \emptyset\}$ is a feasible partition for $\mathcal{H}(\mathcal{M}, \Xi)$. We call $\Gamma$ the coalition structure associated with (or corresponding to) $\pi$. Furthermore, we call $\pi$ $\Xi$-core-stable for $\mathcal{M}$ if $\Gamma$ is core stable for $\mathcal{H}(\mathcal{M}, \Xi)$.

### 3.3  Set Extensions

There are many ways to extend an ordering over the elements of a set $X$ to an ordering over the subsets of $X$ (see Barberá et al. [6] for an extensive survey). Each of these yields a particular way for how to obtain a hedonic game from a FeaCoMP. We are going to use two particular set extensions, both of which are natural and well-known. The "universal responsive set extension" is sometimes just called the "responsive set extension"; what we call the "top-oriented set extension" does not seem to have a consensus name, but it does appear, e.g., in the guise of "lexicographic" utilities or choice functions [3, 10].

**Definition 9 (universal responsive set extension).** *Let $\succsim$ be a weak order over $X$. The universal responsive set extension $\mathfrak{R}$ is defined for all $A, B \subseteq X$ by: $A \succsim^{\mathfrak{R}} B \iff$ there exists an injective mapping $\iota : B \to A$ with $\iota(b) \succsim b$ for all $b \in B$.*

**Definition 10 (top-oriented set extension).** *Let $\succeq$ be a total order over a finite set $X$. Define the scoring vector $s = \left(\frac{1}{2^0}, \frac{1}{2^1}, \frac{1}{2^2}, \ldots, \frac{1}{2^{|X|-1}}\right) = \left(1, \frac{1}{2}, \frac{1}{4}, \ldots, \frac{1}{2^{|X|-1}}\right)$, the rank of $y \in X$ by $r(y,\succeq) = |\{\hat{y} \in X \mid \hat{y} \succ y\}| + 1$, and the value (or utility) of a set $Y \subseteq X$ according to $\succeq$ by $\omega_{\succeq,s}(Y) = \sum_{y \in Y} s_{r(y,\succeq)}$. The top-oriented set extension $\succeq^{top}$ is defined by $\forall A, B \in 2^X : A \succeq^{top} B \Longleftrightarrow \omega_{\succeq,s}(A) \geq \omega_{\succeq,s}(B)$. Note that $\succeq^{top}$ is again a total order.*

**Lemma 1.** *Let $\succeq$ be a total order over a finite set $X$. For all $A, B \subseteq X$ it holds that $A \succ^{top} B \Longleftrightarrow \exists a \in A \setminus B : \forall b \in B \setminus A : a \succ b$.*

**Proof.** First, note that we may reduce to the case where $A$ and $B$ are disjoint, since $\omega_{\succeq,s}(A) > \omega_{\succeq,s}(B)$ holds if and only if $\omega_{\succeq,s}(A \setminus B) = \omega_{\succeq,s}(A) - \omega_{\succeq,s}(A \cap B) > \omega_{\succeq,s}(B) - \omega_{\succeq,s}(A \cap B) = \omega_{\succeq,s}(B \setminus A)$. Now let $a' = \max_{\succeq}(A)$ and $b' = \max_{\succeq}(B)$. If $a' \succ b'$, we have $a \succ b$ for all $b \in B$ and with $i = r(a,\succeq)$ we have that $\omega_{\succeq,s}(A) \geq s_i = 2^{-i+1} \geq \sum_{j=i}^{|X|-1} 2^{-j} \geq \omega_{\succeq,s}(B)$, and so $A \succ B$. By symmetry, $b' \succ a'$ implies $B \succ A$. $\square$

### 3.4  Stability in RAPs, FeaCoMPs, and Hedonic Games

With our interpretation of FeaCoMPs as hedonic games we can now apply stability concepts in hedonic games to FeaCoMPs and RAPs in particular. Since our translation depends on the choice of a set extension rule, we obtain various stability concepts in this way. We will see that some of these correspond precisely to the stability concepts studied by Aziz et al. [4].

We call a quasi-ordering $\succsim$ on the power set $2^X$ of some set $X$ *strictly monotonic* if it holds that $A \succ B$ for all $A, B \in 2^X$ with $A \supsetneq B$. We call a set extension rule $\Xi$ *strictly monotonic* if $\succsim^\Xi$ is strictly monotonic for all quasi-orderings $\succsim$.

**Lemma 2.** *Let $\Xi$ be a set extension rule. Let $\mathcal{M} = (F, L, \succsim, \Phi)$ be a FeaCoMP and $\pi$ a feasible allocation for $\mathcal{M}$. Let $\Gamma$ be the coalition structure of $\mathcal{H}(\mathcal{M},\Xi)$ associated with $\pi$. Then (i) $\pi$ is individually rational for $\mathcal{M}$ if and only if $\Gamma$ is individually rational for $\mathcal{H}(\mathcal{M},\Xi)$ and (ii) if $\Xi$ is strictly monotonic then $\pi$ is individually rational and non-wasteful for $\mathcal{M}$ if and only if $\Gamma$ is individually stable for $\mathcal{H}(\mathcal{M},\Xi)$.*

**Proof.** Statement (i) is clear from our definitions, since every $p \in F \cup L$ strictly prefers the singleton coalition $\{p\}$ to all coalitions containing an inacceptable agent.

This in particular implies that $\Gamma$ is not individually stable if $\pi$ is not individually rational. Assume $\pi$ is wasteful. That means that there exists some pair $(f,l) \in F \times L$ so that $f$, who is acceptable for $l$, strictly prefers $l$ over $\pi(f)$ and $\pi^{-1}(l) \cup \{f\}$ is feasible for $l$. Let $k = \{l\} \cup \pi^{-1}(l)$, so $k \in \Gamma$. Furthermore, $k \cup \{f\} \in \mathcal{F}$, $k \cup \{f\} \succ_f^\star \Gamma(f)$, and for all $f' \in k \cap F$ it holds that $k \cup \{f\} \sim_{f'}^\star k$. Finally, the fact that $f \succsim_l \emptyset$ and the monotonicity of $\Xi$ imply that $k \cup \{f\} \succ_l^\star k$.

Conversely, assume that $\Gamma$ is not individually stable. Then let $p \in F \cup L$ and $k \in \Gamma \cup \{\emptyset\}$ such that $k \cup \{p\} \in \mathcal{F}$, $k \cup \{p\} \succ_p^\star \Gamma(p)$ and $k \cup \{p\} \succsim_{p'}^\star k$ for all $p' \in k$. If $k = \emptyset$, then $\Gamma$ is not individually rational, so by (i), $\pi$ is not individually rational. So we may assume that $k \neq \emptyset$, meaning that $l \in k$ for some $l \in L$. If $k$ contains some family that is inacceptable for $l$, then again $\pi$ is not individually rational, so we may assume that this is not the case. Now consider first the case that $p \in L$. But then, since $k \cup \{p\}$ is feasible, we must have $p = l \in k$. But then $k \cup \{p\} = \Gamma(p)$, contradicting $k \cup \{p\} \succ_p^\star \Gamma(p)$.

Second, consider the case that $p \in F$. Then $k \cup \{p\} \succ_p^\star \Gamma(p)$ implies that or $l \succ_p \pi(p)$. Furthermore, since $\{l\} \cup \pi^{-1}(l) \cup \{p\} = k \cup \{p\} \in \mathcal{F}$, we have that $\pi^{-1}(l) \cup \{p\}$ is feasible for $l$. Finally, the fact that $k \cup \{p\} \succsim_l^\star k$ shows that $p$ is acceptable for $l$. Thus we have proven that $\pi$ is wasteful. $\qquad\square$

Since core stability implies individual stability, the following corollary is an immediate consequence of the above lemma.

**Corollary 1.** *Let $\Xi$ be a strictly monotonic set extension rule. Let $\mathcal{M} = (F, L, \succsim, \Phi)$ be a FeaCoMP and $\pi$ a feasible allocation for $\mathcal{M}$.*

*If $\pi$ is $\Xi$-core stable then $\pi$ is individually rational and non-wasteful.*

**Theorem 1.** *Let $\mathcal{M} = (F, L, \succsim, \Phi)$ be a FeaCoMP without indifferences and $\pi$ a feasible allocation for $\mathcal{M}$. Then $\pi$ is stable if and only if $\pi$ is top-core stable (i.e., $\Xi$-core stable, where $\Xi = $ top is the top-oriented set extension).*

**Proof.** Let $\mathcal{H} = \mathcal{H}(\mathcal{M}, \mathrm{top}) = (F \cup L, \mathcal{F}, \succsim^\star)$ be the hedonic game corresponding to $\mathcal{M}$, using the top-oriented set extension and let $\Gamma$ be the partition corresponding to $\pi$.

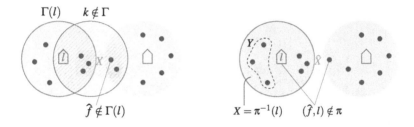

**Fig. 1.** Illustrating the proof of Theorem 1

Assume first that $\Gamma$ is not core stable. We will show that $\pi$ is not stable. Let $k \in \mathcal{F}$ be a blocking coalition for $\Gamma$, i.e., for all $p \in k$ it holds that $k \succsim_p^\star \Gamma(p)$ and there is some player $\hat{p} \in k$ such that $k \succ_{\hat{p}}^\star \Gamma(\hat{p})$. Then $k$ cannot consist of a single family, as that implies that $\pi$ is not individually rational (whence we are done).

So $k = \{l\} \cup X$ for some $l \in L$ and $X \subseteq F$. Also, $\Gamma(l) = \{l\} \cup \pi^{-1}(l)$. Furthermore, it holds that $k \succsim_l^\star \Gamma(l)$. If there exists an $f \in \pi^{-1}(l)$ with $\emptyset \succ_l f$, then $\pi$ is not individually rational and we are done. Otherwise, $X \succsim_l^{\mathrm{top}} \pi^{-1}(l)$. It cannot hold that $X = \pi^{-1}(l)$, since that would mean that $k = \Gamma(l) \in \Gamma$, so no member of $k$ would strictly prefer $k$ to their current coalition. It follows that $X \succ_l^{\mathrm{top}} \pi^{-1}(l)$. By Lemma 1, this means that there is a family $\hat{f} \in X \setminus \pi^{-1}(l)$ such that

$$\hat{f} \succ_l f \qquad \text{for all } f \in \pi^{-1}(l) \setminus X. \tag{1}$$

We now show that $(\hat{f}, l)$ is a blocking pair for $\pi$ (see left side of Fig. 1). First, note that $\hat{f} \in k$, so $k \succsim_{\hat{f}}^\star \Gamma(\hat{f})$, meaning (unless $\pi$ is not individually rational) $l \succsim_{\hat{f}} \pi(\hat{f})$. Since $\hat{f} \notin \pi^{-1}(l)$, we have $l \neq \pi(\hat{f})$, so in fact, $l \succ_{\hat{f}} \pi(\hat{f})$, by the assumption that all the preferences are total. Next, let $Y = \pi^{-1}(l) \setminus X \subseteq \pi^{-1}(l)$. By (1), we have that $\hat{f} \succ_l f$ for all $f \in Y$. Finally, it holds that $(\pi^{-1}(l) \setminus Y) \cup \{\hat{f}\} \subseteq (\pi^{-1}(l) \setminus Y) \cup X = X$. Since $k = \{l\} \cup X \in \mathcal{F}$, the set $X$ is feasible for $l$, and then, a fortiori, so is $(\pi^{-1}(l) \setminus Y) \cup \{\hat{f}\} \subseteq X$. This shows that $(\hat{f}, l)$ is a blocking pair for $\pi$, so $\pi$ is not stable.

Conversely, assume that $\pi$ is not stable. If $\pi$ is not individually rational or wasteful, then Corollary 1 and the fact that the top-oriented set extension is strictly monotonic show that $\Gamma$ is not core stable. So we need only consider the case that there exists a blocking pair $(\hat{f}, l)$ satisfying the following: (i) $l \succ_{\hat{f}} \pi(\hat{f})$, and (ii) there is a set $Y \subseteq \pi^{-1}(l)$ such that $\hat{f} \succ_l f$ for all $f \in Y$ and $(\pi^{-1}(l) \setminus Y) \cup \{\hat{f}\}$ is feasible for $l$ (see right side of Fig. 1). Let $\hat{X} = (\pi^{-1}(l) \setminus Y) \cup \{\hat{f}\}$. Now we show that $\hat{k} = \hat{X} \cup \{l\}$ is a blocking coalition for $\Gamma$. First note that (ii) above guarantees that $\hat{X}$ is feasible for $l$, so $\hat{k} \in \mathcal{F}$.

For all $f \in \pi^{-1}(l)$ it holds by definition that $\hat{k} \sim_f^\star \Gamma(f)$. Furthermore, (i) above implies that $\hat{k} \succ_{\hat{f}}^\star \Gamma(\hat{f})$. Finally, the fact that $\hat{f} \succ_l f$ for all $f \in Y$ implies that $\hat{X} \succ_l^{\text{top}} \pi^{-1}(l)$, by Lemma 1, so $\hat{k} \succ_l^\star \Gamma(l)$. Therefore, $\hat{k}$ is a blocking coalition for $\Gamma$.    □

**Proposition 2.** *Let $\mathcal{M}$ be a FeaCoMP and $\pi$ a feasible allocation for $\mathcal{M}$. If $\pi$ is $\mathfrak{R}$-core stable (i.e., $\Xi$-core stable with $\Xi = \mathfrak{R}$ is the universal responsive set extension), then $\pi$ is weakly stable.*

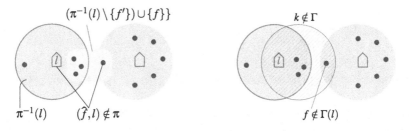

Fig. 2. Illustrating the proof of Proposition 2

**Proof.** Let $\mathcal{H} = \mathcal{H}(\mathcal{M}, \mathfrak{R}) = (F \cup L, \mathcal{F}, \succeq^\star)$ be the hedonic game corresponding to $\mathcal{M}$, using the universal responsive set extension and let $\Gamma$ be the partition corresponding to $\pi$. Assume that $\pi$ is not weakly stable. If $\pi$ is wasteful or not individually rational, then $\pi$ is not $\mathfrak{R}$-core stable, by Corollary 1 and the fact that $\mathfrak{R}$ is strongly monotonic. So, suppose that there exists a *strong blocking pair* $(f, l) \in F \times L$, i.e., that the following holds: (i) $l \succ_f \pi(f)$, and (ii) there is an $f' \in \pi^{-1}(l)$ such that $f \succ_l f'$ and $(\pi^{-1}(l) \setminus \{f'\}) \cup \{f\}$ is feasible for $l$. Now $\Gamma(l) = \{l\} \cup \pi^{-1}(l)$ where $\pi^{-1}(l) \subseteq F$ is a set of families acceptable to $l$. Let

$$k = (\Gamma(l) \setminus \{f'\}) \cup \{f\} = \{l\} \cup ((\pi^{-1}(l) \setminus \{f'\}) \cup \{f\}).$$

We will show that $k$ is a blocking coalition for $\Gamma$ (see Fig. 2). Firstly, we have $f \succ_l f' \succsim_l \emptyset$, hence $k$ contains only families acceptable to $l$. Furthermore, by the definition of the universal responsive set extension, $(\pi^{-1}(l) \setminus \{f'\}) \cup \{f\} \succ_l^{\mathfrak{R}} \pi^{-1}(l)$ and, hence, $k \succ_l^\star \Gamma(l)$. Moreover, $l \succ_f \pi(f)$ implies that $k \succ_f^\star \Gamma(f)$. Finally, for all $\tilde{f} \in \pi^{-1}(l) \setminus \{f'\}$

it holds that $k \sim_{\tilde{f}} \Gamma(\tilde{f})$ (since $k$ and $\Gamma(\tilde{f})$ contain the same locality $l$). This shows that $k$ is a blocking coalition for $\Gamma$. □

However, weak stability does not imply $\Re$-core stability in general, not even for RAPs, as we can see in the following counterexample.

*Example 1 (weak stability $\not\Rightarrow$ $\Re$-core stability).* Let $\mathcal{R} = (F, L, \succeq, S, d, c)$ be a RAP with $F = \{f_1, f_2, f_3, f_4\}$, $L = \{l_1, l_2\}$, $S = \{s_1, s_2\}$, and the preferences, needs, and service capacities are as follows:

$$l_2 \succ_{f_1} l_1 \succ_{f_1} \emptyset, \ d_1 = (1,2), \quad l_2 \succ_{f_2} l_1 \succ_{f_2} \emptyset, \ d_2 = (2,1),$$
$$l_1 \succ_{f_3} l_2 \succ_{f_3} \emptyset, \ d_3 = (3,0), \quad l_1 \succ_{f_4} l_2 \succ_{f_4} \emptyset, \ d_4 = (0,3),$$
$$f_4 \succ_{l_1} f_3 \succ_{l_1} f_2 \succ_{l_1} f_1 \succ_{l_1} \emptyset, \quad f_1 \succ_{l_2} f_2 \succ_{l_2} f_3 \succ_{l_2} f_4 \succ_{l_2} \emptyset, \quad c_1 = c_2 = (3,3).$$

We use the universal set extension rule $\Re$ for translating $\mathcal{R}$ into a hedonic game $\mathcal{H}^{\mathcal{F}} = \mathcal{H}(\mathcal{R}, \Re) = (P, \mathcal{F}, \succeq)$, where

$$P = F \cup L = \{f_1, f_2, f_3, f_4, l_1, l_2\}, \ \succeq = (\succeq_{f_1}^{\star}, \succeq_{f_2}^{\star}, \succeq_{f_3}^{\star}, \succeq_{f_4}^{\star}, \succeq_{l_1}^{\star}, \succeq_{l_2}^{\star}).$$

The modified preference lists of the refugee families $f_i$ with $i \in \{1,2\}$ and $f_j$ with $j \in \{3,4\}$ consist of three indifference classes each:

$$\{f_i, l_2\} \sim_{f_i}^{\star} \cdots \sim_{f_i}^{\star} \{f_1, f_2, f_3, f_4, l_2\} \succ_{f_i}^{\star} \{f_i, l_1\} \sim_{f_i}^{\star} \cdots \sim_{f_i}^{\star} \{f_1, f_2, f_3, f_4, l_1\} \succ_{f_i}^{\star} \{f_i\},$$
$$\{f_j, l_1\} \sim_{f_j}^{\star} \cdots \sim_{f_j}^{\star} \{f_1, f_2, f_3, f_4, l_1\} \succ_{f_j}^{\star} \{f_j, l_2\} \sim_{f_j}^{\star} \cdots \sim_{f_j}^{\star} \{f_1, f_2, f_3, f_4, l_2\} \succ_{f_j}^{\star} \{f_j\}.$$

For the localities, certain coalitions will be incomparable with respect to the extended preferences, but we can visualize these preferences as a Hasse diagram:

$$\succeq_{l_1}^{\star}: \{f_3, f_4, l_1\} \qquad\qquad \succeq_{l_2}^{\star}: \{f_1, f_2, l_2\}$$

$$\{f_4, l_1\} ? \{f_1, f_2, l_1\} \qquad \{f_1, l_2\} ? \{f_3, f_4, l_2\}$$

$$\{f_3, l_1\} \qquad\qquad \{f_2, l_2\}$$

$$\{f_2, l_1\} \qquad\qquad \{f_3, l_2\}$$

$$\{f_1, l_1\} \qquad\qquad \{f_4, l_2\}$$

$$\{l_1\} \qquad\qquad\qquad \{l_2\}$$

The following is a feasible allocation for $\mathcal{R}$: $\pi = \{(f_1, l_1), (f_2, l_1), (f_3, l_2), (f_4, l_2)\}$. The corresponding partition $\Gamma$ for $\mathcal{H}^{\mathcal{F}}$ is $\Gamma = \{k_1, k_2\}$ with $k_1 = \{f_1, f_2, l_1\}$ and $k_2 = \{f_3, f_4, l_2\}$ (see Fig. 3).

It is easy to check that $\pi$ is individually rational (all the localities and families find each other acceptable anyway) and non-wasteful (all capacities are exhausted in $\pi$).

There do exist pairs preferring each other over their partners in $\pi$—namely $(f_1, l_2)$, $(f_2, l_2)$, $(f_3, l_1)$, $(f_4, l_1)$—but none of these are strongly blocking, since no locality can make room for another family by ejecting just one currently assigned family. Hence, $\pi$ is weakly stable.

$\pi:$     $\Gamma:$     $\hat{k} \notin \Gamma$

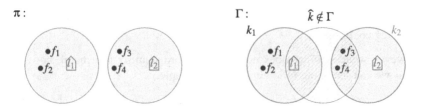

**Fig. 3.** Illustrating Counterexample 1

However, $\Gamma$ is not core stable (and so $\pi$ is not $\mathfrak{R}$-core stable). Indeed, consider the coalition $\{f_3, f_4, l_1\} = \hat{k} \in \mathcal{F}$. Then

1. $\hat{k} \succ^\star_{f_3} k_2 = \Gamma(f_3)$ since $l_1 \succ_{f_3} l_2$,
2. $\hat{k} \succ^\star_{f_4} k_2 = \Gamma(f_4)$ since $l_1 \succ_{f_4} l_2$, and
3. for $X = \Gamma(l_1) \setminus \{l_1\} = \{f_1, f_2)\}$, $\hat{X} = \hat{k} \setminus \{l_1\} = \{f_3, f_4\}$ there exists an injective mapping $\iota \colon X \to \hat{X}$ with $\iota(f) \succsim_{l_1} f$ for all $f \in X$, namely by letting $\iota(f_1) = f_3$ and $\iota(f_2) = f_4$.

Thus $\hat{k}$ is a blocking coalition for $\Gamma$.

**Theorem 2.** *Let $\mathcal{M}$ be a FeaCoMP without indifferences and $\pi$ a feasible allocation. Then $\pi$ is $\mathfrak{R}$-core stable if $\pi$ is stable.*

**Proof.** Let $\mathcal{H} = \mathcal{H}(\mathcal{M}, \mathfrak{R}) = (F \cup L, \mathcal{F}, \succsim^\star)$ be the hedonic game corresponding to a FeaCoMP $\mathcal{M} = (F, L, \succsim, \Phi)$, using the universal responsive set extension and let $\Gamma$ be the partition corresponding to $\pi$. Assume that $\mathcal{M}$ is not core stable. Then there exists a blocking coalition $k \in \mathcal{F}$ for $\Gamma$, i.e., for all $p \in k$ it holds that $k \succsim^\star_p \Gamma(p)$ and there is some player $\hat{p} \in k$ such that $k \succ^\star_{\hat{p}} \Gamma(\hat{p})$. Since $k \in \mathcal{F}$, this coalition either consists of a single refugee family $f$ or it contains exactly one locality. In the first case, it must hold that $f = \hat{p}$. But then $\{f\} = k \succ^\star_f \Gamma(f)$ implies that $\emptyset \succ_f \pi(f)$, so $\pi$ is not individually rational and hence not stable. In the second case, we have $k = \{l\} \cup X$ for some $l \in L$ and $X \subseteq F$ and it holds that $k \succsim^\star_l \Gamma(l)$. Write $\Gamma(l) = \{l\} \cup X'$ with $X' \subseteq F$. If $\Gamma(l)$ contains a family that is inacceptable to $l$, then $\pi$ is not individually rational and hence not stable. Otherwise, the definition of $\succsim^\star_l$ yields that $X \succsim^{\mathfrak{R}}_l X'$. That means there exists an injective mapping $\iota \colon X' \to X$ with $\iota(p) \succsim_l p$ for all $p \in X'$. Without loss of generality, we may assume that $\iota(f) = f$ for all $f \in X' \cap X$ and $\iota(f) \in X \setminus X'$ for all $f \in X' \setminus X$. Since $k$ is blocking for $\Gamma$, we cannot have $k \in \Gamma$. Hence, $k \neq \Gamma(l)$, meaning $X \neq X'$.

If $X' \subseteq X$, then $X \setminus X' \neq \emptyset$, so pick any $f \in X \setminus X'$. Note that $\Gamma(l) \cup \{f\}$ is a feasible coalition, since it is a subset of the feasible coalition $k$. Furthermore, $f \notin X'$ means that $f \notin \Gamma(l)$, so $\pi(f) \neq l$. But $k \succsim^\star_f \Gamma(f)$ implies that $l \succsim_f \pi(f)$. By the assumption that there are no indifferences in the preferences of the RAP $\mathcal{R}$, we have $l \succ_f \pi(f)$, so $\Gamma(l) \cup \{f\} \succ^\star_f \Gamma(f)$. We also have $X' \cup \{f\} \succ^{\mathfrak{R}}_l X'$, so $\Gamma(l) \cup \{f\} \succ^\star_l \Gamma(l)$. For all $f' \in X'$ we have that $\Gamma(l) \cup \{f\} \sim^\star_{f'} \Gamma(l)$. This shows that $\Gamma$ is not individually stable and hence wasteful by Corollary 1.

It remains to consider the case that $X' \setminus X \neq \emptyset$. Let $f'$ be the family most preferred by $l$ among all families in $Y = X' \setminus X$ and let $f = \iota(f') \in X \setminus X'$. We will show that $(f, l)$

is a blocking pair for $\pi$. Firstly, since $k$ is a blocking coalition, $k \succsim_f^\star \Gamma(f)$, meaning that $l \succsim_f \pi(f)$. Since $f \notin X'$, we have $l \neq \pi(f)$ and the assumption that all the preferences of the FeaCoMP $\mathcal{M}$ are strict implies that $l \succ_f \pi(f)$. Next, note that $f = \iota(f') \succsim_l f'$ and since $f \in X$ and $f' \notin X$, we have $f \neq f'$. By the assumption that the preferences of the FeaCoMP $\mathcal{M}$ are strict, we then have $f \succ_l f'$, and by the choice of $f'$ it follows that $f \succ_l \tilde{f}$ for all $\tilde{f} \in Y$. Furthermore, $(\pi^{-1}(l) \setminus Y) \cup \{f\} = (X' \setminus (X' \setminus X)) \cup \{f\} = (X \cap X') \cup \{f\} \subseteq X$, so $(\pi^{-1}(l) \setminus Y) \cup \{f\}$ is feasible for $l$, since $X$ is. This completes the proof that $(f, l)$ is a blocking pair for $\pi$.     $\square$

However, $\mathfrak{R}$-core stability does not imply stability in general, not even for RAPs, as we can see in the following counterexample.

*Example 2 ($\mathfrak{R}$-core stable $\not\Rightarrow$ stable).* Let $\mathcal{R} = (F, L, \succsim, S, d, c)$ be a RAP with four families $F = \{f_1, f_2, f_3, f_4\}$, two localities $L = \{l_1, l_2\}$, a single service $S = \{s_1\}$, and preferences, needs, and service capacities given as follows:

$$l_1 \succ_{f_1} l_2, \ \ l_2 \succ_{f_i} l_1, \ \ \ d_1 = (2), \ d_i = (1) \ \ \text{for } i \in \{2, 3, 4\},$$
$$f_1 \succ_{l_j} f_2 \succ_{l_j} f_3 \succ_{l_j} f_4, \ \ \ c_1 = (3), \ c_2 = (2) \ \ \text{for } j \in \{1, 2\}.$$

We use the universal set extension rule $\Xi = \mathfrak{R}$ for translating $\mathcal{R}$ into a hedonic game $\mathcal{H}^{\mathcal{F}} = \mathcal{H}(\mathcal{R}, \Xi) = (P, \mathcal{F}, \succeq)$:

$$P = F \cup L = \{f_1, f_2, f_3, f_4, l_1, l_2\}, \ \ \ \ \succsim = (\succsim_{f_1}^\star, \succsim_{f_2}^\star, \succsim_{f_3}^\star, \succsim_{f_4}^\star, \succsim_{l_1}^\star, \succsim_{l_2}^\star).$$

The extended preference lists of the refugee families $f_1$ and $f_i$ with $i \in \{2, 3, 4\}$ consist of three indifference classes each:

$$\{f_1, l_1\} \sim_{f_1}^\star \cdots \sim_{f_1}^\star \{f_1, f_2, f_3, f_4, l_1\} \succ_{f_1}^\star \{f_1, l_2\} \sim_{f_1}^\star \cdots \sim_{f_1}^\star \{f_1, f_2, f_3, f_4, l_2\} \succ_{f_1}^\star \{f_1\},$$
$$\{f_i, l_2\} \sim_{f_i}^\star \cdots \sim_{f_i}^\star \{f_1, f_2, f_3, f_4, l_2\} \succ_{f_i}^\star \{f_i, l_2\} \sim_{f_i}^\star \cdots \sim_{f_i}^\star \{f_1, f_2, f_3, f_4, l_2\} \succ_{f_i}^\star \{f_i\}.$$

The Hasse diagrams for the localities' extended preferences look as follows:

$$\succsim_{l_1}^\star : \{f_1, f_2, l_1\} \wr \{f_2, f_3, f_4, l_1\} \qquad\qquad \succsim_{l_2}^\star : \{f_1, l_2\} \wr \{f_2, f_3, l_2\}$$

Let us now consider the outcome $\pi = \{(f_1,l_2),(f_2,l_1),(f_3,l_1),(f_4,l_1)\}$, which is feasible for $\mathcal{R}$, and the corresponding partition $\Gamma = \{k_1,k_2\}$ with $k_1 = \{f_2,f_3,f_4,l_1\}$ and $k_2 = \{f_1,l_2\}$ for $\mathcal{H}^{\mathcal{F}}$ (see Fig. 4). The outcome $\pi$ is not stable. Namely, $(f_1,l_1)$ is a blocking pair, because

(i) $l_1 \succ_{f_1} l_2 = \pi(f_1)$ and
(ii) taking $Y = \{f_2,f_3,f_4\} \subseteq \pi^{-1}(l_1)$ it holds that $\forall f \in Y : f_1 \succ_{l_1} f$ and $(\pi^{-1}(l_1) \setminus Y) \cup \{f_1\} = \{f_1,l_1\}$ is feasible for $l_1$.

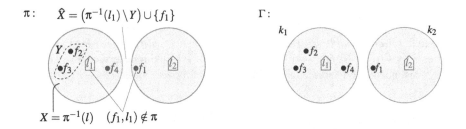

**Fig. 4.** Illustrating Counterexample 2

However, $\Gamma$ is core stable, respectively $\pi$ is $\mathfrak{R}$-*core stable*, because there exists no blocking coalition $\hat{k} \in \mathcal{F}$. To see this, note that such a coalition would have to contain a locality (no family is better off alone than it is in $\pi$). Consider first the case that $\hat{k}$ is a coalition containing $l_1$. For $\hat{k} \succsim_{l_1} \Gamma(l_1) = \{l_1,f_2,f_3,f_4\}$ to hold, there would have to be an injective map from $\{f_2,f_3,f_4\}$ to $\hat{k} \setminus \{l_1\}$. In particular, $\hat{k}$ would have to contain at least three families. But other than $k_1$ there is no feasible coalition containing $l_1$ and at least three families.

Next, consider the case that $\hat{k}$ is a coalition containing $l_2$. For $\hat{k} \succsim_{l_2} \Gamma(l_2) = \{f_1,l_2\}$ to hold, there would have to be an injective map $\iota : \{f_1\} \to \hat{k} \setminus \{l_2\}$ with $\iota(f_1) \succsim_{l_2} f_1$. But since $f_1$ is this locality's preferred family, this would imply that $\iota(f_1) = f_1$, so $f_1 \in \hat{k}$. But $f_1$ already exhausts the capacities of $l_2$, so $\hat{k}$ would have to be $\Gamma(l_2)$, which is not blocking.

## 4   Conclusions

We have proposed a way of transforming the refugee allocation problem as considered by Aziz et al. [4] into the context of hedonic games. This transformation is parametrized by a set extension rule, leading to different variants of individual and core stability. We studied how these notions relate to the various stability concepts introduced in Aziz et al. [4]. Since hedonic games have been studied in much detail and in many variants, this opens the door to various extensions of the refugee allocation model. First of all, there are many more set extension rules (see, e.g., the book chapter by Barberá et al. [6] for an extensive survey) than the two we picked here, reflecting different ways of interpreting the priorities of localities when rating sets of families. Further afield, one might consider looking beyond preferences of localities that are induced by individual-level preferences or incorporating the presence or absence of other families in the preferences of families.

From the viewpoint we elaborated on in this paper, RAPs together with a set extension rule can be considered as a compact representation for a particular, interesting class of hedonic games, so the computational complexity of stability in such games merits studying. Of course, the results by McDermid and Manlove [20] and Aziz et al. [4] on NP-hardness of deciding existence of stable resp. weakly stable allocations are a step in this direction, but it is possible that more general statements can be made for larger classes of set extensions or for other compact representations of FeaCoMPs.

# References

1. Alkan, A.: Non-existence of stable threesome matchings. Math. Soc. Sci. **16**, 207–209 (1988)
2. Aziz, H.: Stable marriage and roommate problems with individual-based stability. In: Proceedings of the 12th International Conference on Autonomous Agents and Multiagent Systems, AAMAS 2013, pp. 287–294. IFAAMAS (2013)
3. Aziz, H., Biró, P., Lang, J., Lesca, J., Monnot, J.: Optimal reallocation under additive and ordinal preferences. In: Proceedings of the 15th International Conference on Autonomous Agents and Multiagent Systems, AAMAS 2016, pp. 402–410. IFAAMAS (2016)
4. Aziz, H., Chen, J., Gaspers, S., Sun, Z.: Stability and pareto optimality in refugee allocation matchings. In: Proceedings of the 17th International Conference on Autonomous Agents and Multiagent Systems, AAMAS 2018, pp. 964–972. IFAAMAS (2018)
5. Aziz, H., Savani, R.: Hedonic games. In: Brandt, F., Conitzer, V., Endriss, U., Lang, J., Procaccia, A. (eds.) Handbook of Computational Social Choice, Chap. 15, pp. 356–376. Cambridge University Press, Cambridge (2016)
6. Barberà, S., Bossert, W., Pattanaik, P.K.: Ranking sets of objects. In: Barberà, S., Hammond, P.J., Seidl, C. (eds.) Handbook of Utility Theory, pp. 893–977. Springer, Boston (2004). https://doi.org/10.1007/978-1-4020-7964-1_4
7. Biró, P., Fleiner, T.: Fractional solutions for capacitated NTU-games, with applications to stable matchings. Discrete Optim. **22**, 241–254 (2016)
8. Biró, P., Fleiner, T., Irving, R.W.: Matching couples with Scarf's algorithm. Ann. Math. Artif. Intell. **77**(3), 303–316 (2016)
9. Biró, P., McDermid, E.: Matching with sizes (or scheduling with processing set restrictions). Discrete Appl. Math. **164**, 61–67 (2014)
10. Chambers, C.P., Yenmez, M.B.: On lexicographic choice. Econ. Lett. **171**, 222–224 (2018)
11. Delacrétaz, D., Kominers, S.D., Teytelboym, A.: Refugee resettlement (2016). Working paper. http://www.t8el.com/jmp.pdf. Version from 8 November 2016
12. Echenique, F., Yenmez, M.B.: A solution to matching with preferences over colleagues. Games Econ. Behav. **59**(1), 46–71 (2007)
13. Elkind, E., Rothe, J.: Cooperative game theory. In: Rothe, J. (ed.) Economics and Computation. An Introduction to Algorithmic Game Theory, Computational Social Choice, and Fair Division, Chap. 3. Springer Texts in Business and Economics, pp. 135–193. Springer, Heidelberg (2015). https://doi.org/10.1007/978-3-662-47904-9_3
14. Gale, D., Shapley, L.S.: College admissions and the stability of marriage. Am. Math. Mon. **69**(1), 9–15 (1962)
15. Goto, M., Kojima, F., Kurata, R., Tamura, A., Yokoo, M.: Designing matching mechanisms under general distributional constraints. Am. Econ. J. Microecon. **9**(2), 226–262 (2017)
16. Hatfield, J.W., Milgrom, P.R.: Matching with contracts. Am. Econ. Rev. **95**(4), 913–935 (2005)

17. Igarashi, A., Elkind, E.: Hedonic games with graph-restricted communication. In: Proceedings of the 15th International Conference on Autonomous Agents and Multiagent Systems, AAMAS 2016, pp. 242–250. IFAAMAS (2016)
18. Irving, R.: An efficient algorithm for the stable roommates problem. J. Algorithms **6**(4), 577–595 (1985)
19. Kamada, Y., Kojima, F.: Efficient matching under distributional constraints: theory and applications. Am. Econ. Rev. **105**(1), 67–99 (2015)
20. McDermid, E.J., Manlove, D.F.: Keeping partners together: algorithmic results for the hospitals/residents problem with couples. J. Comb. Optim. **19**(3), 279–303 (2010)
21. Nguyen, T., Vohra, R.: Near-feasible stable matchings with couples. Am. Econ. Rev. **108**(11), 3154–3169 (2018)
22. Roth, A.E.: Stability and polarization of interests in job matching. Econometrica **52**(1), 47–57 (1984). http://www.jstor.org/stable/1911460
23. Roth, A.E.: Conflict and coincidence of interest in job matching: some new results and open questions. Math. Oper. Res. **10**(3), 379–389 (1985)
24. Roth, A.E.: Deferred acceptance algorithms: history, theory, practice, and open questions. Int. J. Game Theory **36**(3–4), 537–569 (2008)
25. Roth, A.E., Sotomayor, M.: Two-sided matching. Handb. Game Theory Econ. Appl. **1**, 485–541 (1992)
26. Woeginger, G.J.: Core stability in hedonic coalition formation. In: van Emde Boas, P., Groen, F.C.A., Italiano, G.F., Nawrocki, J., Sack, H. (eds.) SOFSEM 2013. LNCS, vol. 7741, pp. 33–50. Springer, Heidelberg (2013). https://doi.org/10.1007/978-3-642-35843-2_4

# Optimizing Social Welfare in Social Networks

Pascal Lange and Jörg Rothe(⊠)

Institut für Informatik, Heinrich-Heine-Universität Düsseldorf, Düsseldorf, Germany
{pascal.lange,rothe}@hhu.de

**Abstract.** We study the computational complexity of envy minimization and maximizing the social welfare of graph-envy-free allocations in social networks. Besides the already known NP-completeness of finding allocations with maximal utilitarian social welfare we prove that NP-completeness is in general also given for the egalitarian social welfare and the Nash product. Moreover, we focus on an extended model, based on directed social relationship graphs and undirected social trading graphs, and analyze the computational complexity of reaching a graph-envy-free allocation by trades with so-called *don't care* agents and without money.

**Keywords:** Fair division · Social welfare optimization · Social network

## 1 Introduction

Studying the computational complexity of finding fair allocations of indivisible resources is the main task in multiagent resource allocation, which is a subtopic of computational social choice (see the book chapters by Lang and Rothe [15] or Bouveret, Chevaleyre and Maudet [6] or the survey by Chevaleyre et al. [8] for an overview).

A common model is given by some resources and a number of agents who are equipped with preference orders or utility functions. A central authority is then asked to compute an allocation that maximizes social welfare. A slightly different setting is the *house market* approach [1] where each agent initially starts with a single resource and is then encouraged to swap their resources under some constraints as long as fairness properties are satisfied. A major drawback of these approaches is the absence of information about relationships between two agents. Social networks, introduced as *negotiation topology* by Chevaleyre et al. [9], are a simple way to describe if two agents are allowed to trade with each other one or to model hierarchies about the agents.

Based on these terms, Gourvès et al. [11] studied a house market model where trades are only allowed between neighbors in an underlying social network. They analyzed the computational complexity of finding a given allocation or checking whether a distinguished resource is reachable by rational trades (that means no trading partner has a disadvantage after the swap) without any kind of money, respecting several undirected graph classes. However, they leave some open questions about the allocation reachability problem that were answered by Bentert et al. [3].

This work was supported in part by DFG grant RO 1202/14-2.

S. Pekeč and K. B. Venable (Eds.): ADT 2019, LNAI 11834, pp. 81–96, 2019.
https://doi.org/10.1007/978-3-030-31489-7_6

Bredereck et al. [7] were interested in the computational complexity of several problems on graph-envy-free allocations in directed social networks. They showed, amongst other things, that it is NP-complete to check whether there is a graph-envy-free allocation that maximizes utilitarian social welfare in acyclic social networks. Several other fairness criteria such as so-called epistemic envy-freeness were studied on social networks by Aziz et al. [2]. Unlike previous papers that use agents as vertices in social networks, Bouveret et al. [5] address networks on resources instead of agents.

We show that the problem of maximizing social welfare for graph-envy-free allocations is also hard for other social welfare measures, including egalitarian social welfare and the Nash product. We also look at so-called *don't care* agents that are allowed to defy the rational constraint of trades but are relevant to graph-envy-freeness, and analyze how hard it is to find a graph-envy-free allocation by rational trades with at most one don't care agent. This setting is based on the one studied by Gourvès et al. [11] but uses a directed social network that does not necessarily need to be the same as the undirected one.

The structure of this paper is as follows: In Sect. 2, we briefly describe the basic model and its extension to social networks. Section 3 is concerned with the computational complexity of finding graph-envy-free allocations with maximal social welfare. Finally, we show in Sect. 4 that it is NP-hard to find a graph-envy-free allocation by rational trades if we allow at least one so-called *don't care* agent. We conclude in Sect. 5 and outline some directions of future research.

## 2 Preliminaries

Let $A = \{a_1, \ldots, a_n\}$ be a finite set of *agents* and $R = \{r_1, \ldots, r_m\}$ a finite set of *resources*. A *preference profile* $\succ = (\succ_1, \ldots, \succ_n)$ *for A and R* consists of $n$ linear orders (i.e., transitive trichotomous binary relations), where $\succ_i$ expresses agent $a_i$'s preferences over $2^R$. When clear from context, we sometimes omit the subscript $i$ from $\succ_i$. An *ordinal allocation setting* is defined as a triple $(A, R, \succ)$.

A *cardinal allocation setting* $(A, R, U)$ is defined analogously, except that the profile $\succ$ of linear orders is replaced by a profile $U = (u_1, \ldots, u_n)$ of *utility functions* $u_i : 2^R \to \mathbb{Q}$, where $u_i(B)$ denotes agent $a_i$'s utility from receiving the bundle $B \subseteq R$ of resources and $\mathbb{Q}$ is the set of rational numbers (though we could use any ordered set instead). For notational convenience, we will also write $u(a_i, \cdot) = u_i(\cdot)$, i.e., all $n$ utility functions are collected in a single two-ary function $u : A \times 2^R \to \mathbb{Q}$. A utility function $u$ is *additive* if it can be written as

$$u(a, B) = \sum_{r \in B} u(a, \{r\})$$

for every agent $a \in A$ and every bundle $B \subseteq R$ of resources. If $u(a, \{r\}) \in \{0, 1\}$ holds for all agents $a \in A$ and every resource $r \in R$, we call $u$ a *binary* utility function.

An *allocation* is defined as a map $\pi : A \to 2^R$ with $\pi(a_i) \cap \pi(a_j) = \emptyset$ for any $a_i, a_j \in A$ with $a_i \neq a_j$, i.e., $\pi(a_i)$ is the bundle of resources assigned to agent $a_i$ under $\pi$, and we assume that resources cannot be shared among several agents. We also require *complete* allocations, i.e., $R = \bigcup_{a_i \in A} \pi(a_i)$, and we denote by $\Pi_{A,R}$ the set of all possible complete allocations.

To evaluate and compare the social welfare of allocations, we make use of *(Bergson–Samuelson) social welfare functions* $sw : \Pi_{A,R} \to \mathbb{Q}$ (see [4]): $sw(\pi)$ gives the *social welfare of allocation* $\pi$. The following three concrete social welfare functions are the most important ones:

1. *utilitarian social welfare*, defined by $sw_u(\pi) = \sum_{a_i \in A} u(a_i, \pi(a_i))$, gives the sum of individual utilities of all agents;

2. *egalitarian social welfare*, defined by $sw_e(\pi) = \min_{a_i \in A} (u(a_i, \pi(a_i)))$, gives the utility of the agent who is worst off (and therefore can also be seen as some measure of fairness); and

3. *Nash product social welfare*, defined by $sw_N(\pi) = \prod_{a_i \in A} u(a_i, \pi(a_i))$, gives the product of individual utilities of all agents, which can be seen as some sort of compromise between utilitarian and egalitarian social welfare because it is monotonic like $sw_u$ and "balanced" like $sw_e$.

A *social trading network* is an undirected graph $G_T = (A, E_T)$, where vertices are identified with agents and an edge $e \in E_T$ represents an allowed way for the two incident agents to *trade* the resources currently allocated to them. The expansion $((A, R, \succ), G_T)$ of an ordinal allocation setting $(A, R, \succ)$ to a social trading network $G_T = (A, E_T)$ is called an *extended ordinal allocation setting*.

A directed graph $G_R = (A, E_R)$ is called a *social relationship network*. In a similar way, we define *extended cardinal allocation settings* as $((A, R, U), G_R)$ with a social relationship network $G_R$.

**Definition 1 (trade).** *Let $(A, R, \succ)$ be an ordinal allocation setting and $a_i, a_j \in A$ be two agents. A* trade *between $a_i$ and $a_j$ is a pair $(\pi, \pi')$ such that $\pi, \pi' \in \Pi_{A,R}$ are two allocations for which the following statements hold:*

1. $\pi(a_i) = \pi'(a_j)$ and $\pi(a_j) = \pi'(a_i)$
2. $\pi(a_\ell) = \pi'(a_\ell)$ for all $a_\ell \in A \setminus \{a_i, a_j\}$.

We will also make use of the more explicit notation $((a_i, \pi(a_i)), (a_j, \pi(a_j)))$ from which we can derive the resulting allocation $\pi'$. We will restrict trades on allocations $\pi, \pi'$ with $|\pi(a_i)| = |\pi'(a_i)| = 1$ for all agents $a_i$.

Two properties a trade is required to satisfy are *validity* and *rationality*.

**Definition 2.** *A trade $(\pi, \pi')$ between $a_i$ and $a_j$ is* valid *if $a_i$ and $a_j$ are neighbors in $G_T$. It is* rational *if $\pi'(a_i) \succ_i \pi(a_i)$ and $\pi'(a_j) \succ_j \pi(a_j)$, i.e., both agents prefer the new bundle they receive under $\pi'$ to the old one assigned to them under $\pi$.*

We now define envy-freeness in a social relationship network.

**Definition 3 (graph-envy-free allocation [7,9]).** *Let $A$ be a set of agents, $G_R = (A, E_R)$ a social relationship network, and $((A, R, U), G_R)$ an extended cardinal allocation setting.*

*1. An allocation $\pi$ is (weakly) graph-envy-free if for two agents $a_i, a_j \in A$, where $a_j$ is a successor of $a_i$ in $G_R$, the following holds: $u(a_i, \pi(a_i)) \geq u(a_i, \pi(a_j))$.*

*2. If the above inequality is strict, $\pi$ is a strong graph-envy-free allocation.*

We assume that the reader is familiar with the basic notions of computational complexity such as the complexity classes P and NP, polynomial-time many-one reducibility, and the related notions of NP-hardness and NP-completeness. A detailed introduction and more background can be found, e.g., in the textbooks by Papadimitriou [17], Rothe [18], and Garey and Johnson [10].

## 3   Social Welfare on Graph-Envy-Free Allocations

In this section we are interested in the computational complexity of (the decision problems associated with) optimizing social welfare of graph-envy-free allocations. First, we define one of the decision problems that we will study.

---

**GEF-NASH-PRODUCT-SOCIAL-WELFARE-OPTIMIZATION (GEF-NSWO)**

**Given:**  An extended cardinal allocation setting $((A, R, U), G)$ with a social relationship network $G$ (i.e., $G$ is a directed graph) and a natural number $K$.

**Question:** Does there exist a graph-envy-free allocation $\pi$ with $sw_N(\pi) \geq K$?

---

By replacing the social welfare function $sw_N$ above by $sw_u$ and $sw_e$, respectively, we get the problems GEF-USWO and GEF-ESWO. GEF-USWO is known to be NP-complete for directed acyclic graphs and three-valued utility functions [7, Proposition 4] but is in P for binary utility functions and acyclic social relationship networks [7, Proposition 5]. We show that NP-completeness also holds for egalitarian social welfare and directed graphs restricted to acyclic graphs. Under weaker assumptions about utility functions and allocations, we can also show that the result holds for the Nash product. For both results, we make use of the well-known strongly NP-complete problem EXACT-COVER-BY-3-SETS [14].[1]

---

**EXACT-COVER-BY-3-SETS (X3C)**

**Given:**  A set $B = \{b_1, \ldots, b_{3m}\}$ for a natural number $m$ and a collection $S = \{S_1, \ldots, S_n\}$ of subsets $S_i \subseteq B$ with $|S_i| = 3$ for each $i$, $1 \leq i \leq n$.

**Question:** Does there exist an exact cover of $B$, i.e., does there exist an index set $I \subseteq \{1, \ldots, n\}$ with $|I| = m$ such that $B = \bigcup_{i \in I} S_i$ is a disjoint union?

---

Our reductions are based on special subgraphs $C_i$ (called *3-center groups*) for every set $S_i \in S$, where the three elements in $S_i$ are represented by three *center nodes*.

---

[1] Garey and Johnson [10] define a problem to be *strongly* NP-*complete* if it is NP-complete even when each of its numerical parameters is bounded by a polynomial in the length of the input. That implies that, unless P = NP, strongly NP-complete problems cannot have fully polynomial-time approximation schemes nor pseudo-polynomial-time algorithms.

**Definition 4 (3-center group).** *Let $V = \{a_1, a_2, a_3, d, y\}$ be a set of five agents. A 3-center group is a directed graph $C = (V, E_C)$ with edges*

$$E_C = \{(y, a_i), (a_i, d) \mid 1 \leq i \leq 3\}.$$

*Agent d is called the* decision maker *and $a_1$, $a_2$, and $a_3$ are called the* center agents.

The main goal then is the construction of utility values for every agent such that the decision maker can be used to decide for $i \in I$ or $i \notin I$. If she decides for $i \notin I$, her three predecessor agents must be allocated resources that are not constructed from the base elements in $B$. For $i \in I$, the center agents can be allocated only resources from $B \cap S_i$.

**Theorem 1.** GEF-ESWO *is strongly* NP-*complete even if the social network is acyclic, the utility values are restricted to $\{0, 1, 2\}$ and $|A| = |R|$ holds.*

**Proof.** Membership of GEF-ESWO in NP is easy to see. To prove NP-hardness, we provide a reduction from X3C. Let $(B, S)$ with $B = \{b_1, \ldots, b_{3m}\}$ and $S = \{S_1, \ldots, S_n\}$ with $S_i = \{S(i, 1), S(i, 2), S(i, 3)\}$ be an instance of X3C. Construct a GEF-ESWO instance $((A, R, U), G, K)$ with $A = \{a^{\mathrm{top}}\} \cup \{a_{(i,j)} \mid 1 \leq i \leq n \text{ and } 1 \leq j \leq 3\} \cup \{x_i \mid 1 \leq i \leq n\} \cup \{y_i \mid 1 \leq i \leq n\}$ and $R = \{r^{\mathrm{top}}\} \cup R^B \cup R^Y \cup R^D \cup R^T \cup R^F$, where $R$ is given by

$$R^B = \{r_j^B \mid 1 \leq j \leq 3m\}, \qquad R^D = \{r_j^D \mid 1 \leq j \leq 3n - 3m\}, \qquad R^T = \{r_i^T \mid 1 \leq i \leq m\},$$

$$R^F = \{r_i^F \mid 1 \leq i \leq n - m\}, \quad \text{and} \quad R^Y = \{r_i^Y \mid 1 \leq i \leq n\}.$$

The additive utility functions are defined by

$$\alpha_{a_{(i,j)}}^{\{r_\ell^B\}} = 1 \qquad (1 \leq i \leq n \text{ and } 1 \leq j \leq 3 \text{ and } b_\ell \in S_i),$$

$$\alpha_{a_{(i,j)}}^{\{r\}} = 1 \qquad (1 \leq i \leq n \text{ and } 1 \leq j \leq 3 \text{ and } r \in R^T),$$

$$\alpha_{a_{(i,j)}}^{\{r\}} = 2 \qquad (1 \leq i \leq n \text{ and } 1 \leq j \leq 3 \text{ and } r \in R^D \cup R^F),$$

$$\alpha_{y_i}^{\{r\}} = 2 \qquad (1 \leq i \leq n \text{ and } r \in R^Y),$$

$$\alpha_{x_i}^{\{r\}} = 1 \qquad (1 \leq i \leq n \text{ and } r \in R^T \cup R^F),$$

$$\alpha_{a^{\mathrm{top}}}^{\{r^{\mathrm{top}}\}} = 2,$$

and 0 otherwise. Set $K = 1$ as a threshold and define the directed social relationship network $G = (A, E)$ with edge set

$$E = \{(a^{\mathrm{top}}, y_i) \mid 1 \leq i \leq n\} \cup$$
$$\{(y_i, a_{(i,j)}), (a_{(i,j)}, x_i) \mid 1 \leq i \leq n \text{ and } 1 \leq j \leq 3\}.$$

(see Fig. 1).

Note that, due to $u(a, \emptyset) = 0$ for all $a \in A$ and $|A| = 5n + 1 = |R|$, exactly one resource must be allocated to every agent. Since the transformation, as described above, can be done in polynomial time, it remains to show that $(B, S)$ is a yes-instance of X3C if and only if $((A, R, U), G, K)$ is a yes-instance of GEF-ESWO.

From left to right, suppose that $(B, S)$ is a yes-instance of X3C. Thus there is an index set $I \subseteq \{1, \ldots, n\}$ with $|I| = m$ such that $B = \bigcup_{i \in I} S_i$ is a disjoint union. Let $\bar{I} = \{1, \ldots, n\} \setminus I$ and define two bijective functions

$$\mu_1 : I \rightarrow \{1, \ldots, m\} \qquad \text{and} \qquad \mu_2 : \bar{I} \rightarrow \{1, \ldots, n - m\}.$$

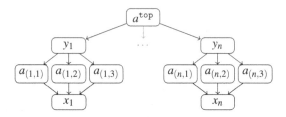

**Fig. 1.** The social relationship network for the proof of Theorem 1 with $n$ 3-center groups

Define the allocation:

$$\pi(a) = \begin{cases} \{r^{\text{top}}\} & \text{if } a = a^{\text{top}} \\ \{r_i^Y\} & \text{if } a = y_i \text{ for an } 1 \leq i \leq n \\ \{r_{S_{(i,j)}}^B\} & \text{if } a = a_{(i,j)} \text{ and } i \in I \\ \{r_z^D\} & \text{if } a = a_{(i,j)} \text{ and } i \in \bar{I} \text{ (for } 1 \leq z \leq 3(n-m)) \\ \{r_{\mu_1(i)}^T\} & \text{if } a = x_i \text{ and } i \in I \\ \{r_{\mu_2(i)}^F\} & \text{if } a = x_i \text{ and } i \in \bar{I} \end{cases}$$

such that every resource is allocated to exactly one agent. Since every agent $a$ gets a resource of value at least 1, we have $u(a, \pi(a)) \geq 1$, and it follows that $sw_e(\pi) \geq K = 1$.

We show that this allocation is graph-envy-free. Due to $u(a^{\text{top}}, \pi(a^{\text{top}})) = 2$ and the fact that all utility values are at most 2, we do not have further restrictions on any child node. Since every agent $y_i$ gets a utility value of 2, there again are no further restrictions and, moreover, there is no conflict with their parent node $a^{\text{top}}$. The agents $a_{(i,j)}$ either receive a resource from $R^B$ or from $R^D$, which means that the utility values are in $\{1, 2\}$:

1. For $i \in I$ and $a \in \{a_{(i,1)}, a_{(i,2)}, a_{(i,3)}\}$, we have $u(a, \pi(a)) = 1$. Their single child $x_i$ gets a resource from $R^T$, which yields utility values of 1.
2. For $i \in \bar{I}$ and $a \in \{a_{(i,1)}, a_{(i,2)}, a_{(i,3)}\}$, we have $u(a, \pi(a)) = 2$. Their single child $x_i$ gets a resource from $R^F$, which yields utility values of 2.

From right to left, suppose that $((A, R, U), G, K)$ is a yes-instance of GEF-ESWO. Then there exists a graph-envy-free allocation $\pi$ with $sw_e(\pi) \geq K = 1$.

First, observe that $u(a, \emptyset) = 0$ holds for every agent $a \in A$. That means every agent must be allocated at least one resource.

Since $u(a^{\text{top}}, r) = 1$ only holds for $r = r^{\text{top}}$, we conclude that $\pi(a^{\text{top}}) = r^{\text{top}}$ is another necessary condition. Also note that agents $x_i$ have a utility value greater than 0 only if they receive items from $R^T \cup R^F$.

It is clear that the three center nodes in a 3-center-group $\{x_i, y_i, a_{(i,1)}, a_{(i,2)}, a_{(i,3)}\}$ must allocate a utility value of 2 if their leaf $x_i$ receive a resource from $R^F$. Due to $|R^D| = 3n - 3m$ and $|R^F| = n - m$, it follows that all resources from $R^D$ must be distributed between these center nodes. So it remains to focus on the remaining $m$ 3-center-groups. As all resources from $R^D$ were allocated as stated above, we have to distribute all resources from $R^B$ to the center agents in 3-center groups who are not allocated any resource yet. Since $u(a_{(i,j)}, r_p) = 1$ if and only if $p \in S_i$, there must be an exact cover, which completes the proof.                                                                                                □

GEF-ESWO is NP-complete for three-valued functions. Therefore, in general GEF-$j$-RANK-DICTATOR, which is defined as follows:[2]

| GEF-$j$-RANK-DICTATOR |
|---|
| **Given:**      An extended cardinal allocation setting $((A, R, U), G)$ with a social relationship network $G$ (i.e., $G$ is a directed graph) and a natural number $K$. |
| **Question:** Does there exist a graph-envy-free allocation $\pi$ with $v_j^*(\pi) \geq K$? |

is also an NP-complete problem. Heinen et al. [12, 13] studied the corresponding problem $j$-RANK-DICTATOR where we don't have a social relationship network and ask simply for allocations (instead of graph-envy-free allocations) $\pi$ such that $v_j^*(\pi)$ meets or exceeds a given value $K$.

The reduction from the proof of Theorem 1 with a $K$ exponential in $n$ can be used to prove NP-completeness of GEF-NSWO for three-valued utility functions restricted to integer numbers as well. However, we can even show *strong* NP-completeness for the Nash product if we allow rational utility values. To this end, we make use of a consequence of the inequality of the arithmetic mean and the geometric mean, which is stated in the following lemma.

**Lemma 1.** *Let $x_1, \ldots, x_n$ be nonnegative numbers satisfying $\sum_{i=1}^n x_i \leq n$. Then $\prod_{i=1}^n x_i \leq 1$, where equality holds if and only if $x_1 = \cdots = x_n = 1$.*

The proof of Theorem 2 itself is based again on a reduction from X3C.

**Theorem 2.** GEF-NSWO *is strongly* NP-*complete even if the underlying social relationship network is acyclic and $|A| = |R|$ holds.*

**Proof.** Membership of GEF-NSWO in NP is again obvious. For proving NP-hardness, we reuse the same idea as for GEF-ESWO. Let $(B, S)$ be a given instance of X3C with $B = \{b_1, \ldots, b_{3m}\}$ and $S = \{S_1, \ldots, S_n\}$ and construct a GEF-NSWO instance with

$$A = \{a^{\text{top}}\} \cup \{a_{(i,j)} \mid 1 \leq i \leq n \text{ and } 1 \leq j \leq 3\} \cup \{x_i \mid 1 \leq i \leq n\} \cup \{y_i \mid 1 \leq i \leq n\}$$
$$R = \{r^{\text{top}}\} \cup R^Y \cup R^B \cup R^D \cup R^T \cup R^F,$$

---

[2] We need some notation to define this problem. Let $v(\pi) = (u_i(\pi_i))_{1 \leq i \leq n}$ be the utility vector induced by an allocation $\pi$. Let $v^*(\pi)$ be the vector that results from $v(\pi)$ by sorting all entries nondecreasingly. In particular, $v_1^*(\pi)$ is the utility of a worst-off agent and thus expresses egalitarian social welfare; $v_{\lceil n/2 \rceil}^*(\pi)$ is the utility of a median-off agent; and $v_n^*(\pi)$ is the utility of a best-off agent and thus expresses so-called elitist social welfare.

where the resource set is given by

$$R^B = \{r_j^B \mid 1 \le j \le 3m\}, \qquad R^D = \{r_j^D \mid 1 \le j \le 3n - 3m\}, \qquad R^T = \{r_i^T \mid 1 \le i \le m\},$$
$$R^F = \{r_i^F \mid 1 \le i \le n - m\}, \quad \text{and} \quad R^Y = \{r_i^Y \mid 1 \le i \le n\}.$$

Note that $|A| = 5n + 1 = |R|$ and construct the additive utility functions, defined by the following coefficients:

$$\alpha_{a_{(i,j)}}^{\{r_\ell^B\}} = 1 \qquad (1 \le i \le n \text{ and } 1 \le j \le 3 \text{ and } b_\ell \in S_i),$$

$$\alpha_{a_{(i,j)}}^{\{r\}} = 1 \qquad (1 \le i \le n \text{ and } 1 \le j \le 3 \text{ and } r \in R^T),$$

$$\alpha_{a_{(i,j)}}^{\{r\}} = 2 \qquad (1 \le i \le n \text{ and } 1 \le j \le 3 \text{ and } r \in R^D \cup R^F),$$

$$\alpha_{x_i}^{\{r\}} = 1 \qquad (1 \le i \le n \text{ and } r \in R^T),$$

$$\alpha_{x_i}^{\{r\}} = \frac{1}{8} \qquad (1 \le i \le n \text{ and } r \in R^F),$$

$$\alpha_{y_i}^{\{r\}} = 1 \qquad (1 \le i \le n \text{ and } r \in R^Y),$$

$$\alpha_{a^{\text{top}}}^{\{r^{\text{top}}\}} = 1$$

and 0 otherwise. Let $G = (A, E)$ with

$$E = \{(a^{\text{top}}, y_i) \mid 1 \le i \le n\} \cup \{(y_i, a_{(i,j)}), (a_{(i,j)}, x_i) \mid 1 \le i \le n \text{ and } 1 \le j \le 3\}.$$

be the directed social relationship network and set $K = 1$. The transformation, as described above, can be done in polynomial time. So it remains to show that $(B, \mathcal{S})$ is a yes-instance of X3C if and only if $((A, R, U), G, K)$ is a yes-instance of GEF-NSWO.

From left to right, suppose that $(B, \mathcal{S})$ is a yes-instance of X3C. Thus there is an index set $I \subseteq \{1, \dots, n\}$ with $|I| = m$ such that $B = \bigcup_{i \in I} S_i$ is a disjoint union. By using the same allocation $\pi$ as in the proof for GEF-ESWO, we get a graph-envy-free allocation and the Nash product reaches the threshold:

$$sw_N(\pi) = \cdot 1^{1+m+n} \cdot 1^{3m} \cdot 2^{3n-3m} \cdot \left(\frac{1}{8}\right)^{n-m} = 2^{3(n-m)} \cdot 2^{-3(n-m)} = 1 \ge K.$$

From right to left, suppose that $(B, \mathcal{S})$ is a no-instance of X3C. Thus there is no index subset $I \subseteq \{1, \dots, n\}$ such that $B = \bigcup_{i \in I} S_i$ with $|S_i| = 3$ is a disjoint union. Let $\hat{I}$ with $|\hat{I}| < m$ be a largest index subset such that $\bigcup_{i \in \hat{I}} S_i$ is a disjoint union. We try to reach the optimal Nash product and show that we will never reach the threshold $K = 1$.

It is easy to see that $\pi(a^{\text{top}}) = r^{\text{top}}$ and $\pi(y_i) \in R^Y$ are necessary for preventing $sw_N(\pi) = 0$. Since every $x_i$ only generates a utility value greater than zero if they get resources from $R^T \cup R^F$ and $|R^T \cup R^F| = n$ holds, it is clear that these resources must be distributed among these agents. We have $n - m$ resources in $R^F$. Distribute these $n - m$ resources among $n - m$ agents $x_i$ with $i \notin \hat{I}$ and collect their indices in a set $I'$. All predecessors $a_{(i,1)}, a_{(i,2)}, a_{(i,3)}$ ($i \in I'$) have to receive items from $R^D$ so as to prevent graph-envy. After that, there is no resource left in $R^D$, due to $|R^D| = 3(n - m)$. By now we have

$$sw_N(\pi) = 1^{m+n+1} \cdot 2^{3n-3m} \cdot \left(\frac{1}{8}\right)^{n-m} \cdot \prod_{i \in \{1,\ldots,n\} \setminus I'} \prod_{j=1}^{3} u(a_{(i,j)}, \pi(a_{(i,j)})).$$

We show that the last product never reaches 1. First note that only resources from $R^B$ are currently unallocated. Collect all agents occurring in this product in a set $\hat{A}$ with $|\hat{A}| = 3m$. By using $\sum_{i \in \hat{A}} u(\hat{a}_i, \pi(\hat{a}_i)) \leq |\hat{A}|$ for $\pi(\hat{a}_i) \in R^B$ and Lemma 1, it follows that $\prod_{i \in \hat{A}} u(\hat{a}_i, \pi(\hat{a}_i)) \leq 1$, where equality holds only if all utility values are 1. Due to the construction of the utility functions and $|\hat{I}| < m$, there remains at least one center agent that cannot generate a utility value of 1: Assume, for the sake of contradiction, that each $a \in \hat{A}$ generates a utility value of 1 if only resources from $R^B$ are available. Then there must be a partition of $R^B$ into $m$ disjoint subsets with cardinality 3, since resources are assumed to be nonshareable. This, however, contradicts $|\hat{I}| < m$.  □

For proving strong NP-hardness of GEF-ESWO and GEF-NSWO, we needed at least two utility values greater than zero to enable the decision maker to make a decision. We now show that both problems are solvable in polynomial time if the utility functions are binary.

---

**Algorithm 1.** GEF-ESWO and GEF-NSWO for binary utility functions

---

**Require:** Extended cardinal allocation setting $((A, R, U), G_R)$ with $|A| = |R|$ normalized additive binary utility functions and directed social relationship network $G_R$
1: Create a bipartite graph $G_B = (A \cup R, E_B)$ with $\{a, r\} \in E_B$ iff $u(a, \{r\}) = 1$
2: $M := \text{MAXIMUMMATCHING}(G_B)$
3: $U := 0$
4: **if** $|M| = |R|$ **then**
5:     $U := 1$
6: **end if**
7: **return** $U$

---

**Theorem 3.** GEF-ESWO *and* GEF-NSWO *are in* P *for binary utility functions, acyclic social relationship networks, and* $|A| = |R|$.

**Proof.** Since all arguments in this proof are true for both egalitarian social welfare and the Nash product, we only give a proof for the Nash product and show that Algorithm 1 solves GEF-NSWO in polynomial time. First, note the following equivalence: There is an allocation $\pi$ with $u(a, \pi(a)) = 1$ for all agents $a \in A$ if and only if there is an allocation $\pi'$ with $sw_N(\pi') = 1$.

If $M$ is a perfect matching (i.e., $2|M| = |A \cup R|$), an allocation $\pi$ with $u(a, \pi(a)) = 1$ for all agents $a \in A$ can be derived from $M$. So the maximal Nash product is 1. Obviously, this allocation is graph-envy-free. If no perfect matching exists, at least one agent remains without any resource or gets a resource $r$ with $u(a, r) = 0$. So only allocations with $sw_N(\pi) \neq 1$ are possible. Since $|A| = |R|$ and all utility functions are normalized (i.e., $\alpha_a^0 = 0$ for all agents $a \in A$), $sw_N(\pi) \geq 2$ is not possible (independent of whether graph-envy-freeness is fulfilled or not), for otherwise we would need at least $|A| + 1$

resources. That means we only can get allocations $\pi$ with $sw_N(\pi) = 0$. As it does not matter how the allocation is constructed, we can set $\pi(a) = R$ for an agent without incoming edges and $\pi(a') = \emptyset$ for all other agents $a' \neq a$, provided that the social relationship network is acyclic. Finding a maximal matching and checking if a graph is acyclic can be done in polynomial time, so our algorithm runs in polynomial time. $\quad\Box$

## 4  Graph-Envy-Free Allocations Through Rational Trades

It is easy to see that there are allocation settings that cannot reach a graph-envy-free allocation. In this section, we analyze the effects of allowing *don't care* agents, distinguished agents in $A$ who are not forced to do rational trades (though they still may be subject to graph envy).

---

**DONT-CARE-AGENT-TRADE-BASED-GEF-ALLOCATION**

---

**Given:**  An ordinal allocation setting $(A, R, \succ)$ with an (undirected) social trading network $G_T = (A, E_T)$, a (directed) social relationship network $G_R = (A, E_R)$, an initial allocation $\pi_0$, and a natural number $K$.

**Question:** Does there exist a graph-envy-free allocation $\pi$ that is reachable from $\pi_0$ with at most $K$ *don't care* agents, i.e., agents who are not forced to do rational trades?

---

As described in the previous section, a social relationship network $G_R$ represents a hierarchy of agents, whereas social trading networks $G_T$ encode which agents are trusted trading partners. We first observe that $G_T = G_R'$ (where $G_R'$ is the undirected adaption of $G_R$) will be a rather strict assumption that some situations cannot be modeled with: Consider a group of two teams with two group leaders, $L_1$ and $L_2$, and two members, $M_1$ and $M_2$, where the hierarchy is encoded in a directed graph with edges from $L_i$ to $M_i$ ($i \in \{1, 2\}$). Then $L_i$ envies member $M_i$ if $M_i$ gets a resource that $L_i$ finds better than her initial one. Moreover, assume that only workers and only leaders are allowed do any trade among themselves. If the social trading network would be the undirected adaption of the social relationship network, then $L_i$ either trades with $M_i$ or cannot envy $M_i$.

**Theorem 4.** DONT-CARE-AGENT-TRADE-BASED-GEF-ALLOCATION *is an* NP-*hard problem, even if the underlying social trading network is a star.*

**Proof.** For proving NP-hardness, we again provide a reduction from the well-known strongly NP-complete problem X3C. Let $(B, S)$ be an instance of X3C with $B = \{b_1, \ldots, b_{3m}\}$ and $S = \{S_1, \ldots, S_n\}$, where each element in $S_i$ is denoted by $S(i, j)$ for $1 \leq j \leq 3$. Construct a DONT-CARE-AGENT-TRADE-BASED-GEF-ALLOCATION instance $((A, R, \succ), G_T, G_R, \pi_0, K)$ with $A = A^0 \cup A^C \cup A_1^D \cup A_2^D \cup A^V \cup A^W \cup A^X \cup A^Y \cup A^Z$ and $R = R^0 \cup R^B \cup R^G \cup R_1^D \cup R_2^D \cup R^E \cup R^C \cup R^N \cup R^S \cup R^T$, where

$$A^0 = \{a_{i,j} \mid 1 \leq i \leq n \text{ and } 1 \leq j \leq 3\}, \quad A^C = \{c_i \mid 1 \leq i \leq n\},$$
$$A_1^D = \{d_i^1 \mid 0 \leq i \leq n\}, \quad A_2^D = \{d_i^2 \mid 1 \leq i \leq 3n - 1\},$$
$$A^W = \{w_i \mid 1 \leq i \leq n\}, \quad A^X = \{x_i \mid 1 \leq i \leq n\},$$
$$A^Y = \{y_i \mid 1 \leq i \leq n\}, \quad A^V = \{v_i \mid 1 \leq i \leq 3n - 3m\}, \text{ and}$$
$$A^Z = \{z_i \mid 1 \leq i \leq 3m\},$$

and

$$R^0 = \{\rho_{i,j} \mid 1 \leq i \leq n \text{ and } 1 \leq j \leq 3\}, \quad R^B = \{\beta_i \mid 1 \leq i \leq 3m\},$$
$$R^G = \{\gamma_i \mid 1 \leq i \leq 3n - 3m\}, \quad R_1^D = \{\delta_i^1 \mid 0 \leq i \leq n - 1\},$$
$$R_2^D = \{\delta_i^2 \mid 0 \leq i \leq 3n - 1\}, \quad R^C = \{\chi_i \mid 1 \leq i \leq n\},$$
$$R^N = \{\nu_i \mid 1 \leq i \leq n\}, \quad R^S = \{\sigma_i \mid 1 \leq i \leq n\},$$
$$R^T = \{\tau_i \mid 1 \leq i \leq m\}, \text{ and} \quad R^E = \{\varepsilon_i \mid 1 \leq i \leq n - m\},$$

and the ordinal preference profiles are defined as follows (omitting the subscripts of $\succ$ for readability):

$$z_q : \delta_{q-1}^2 \succ \boxed{\beta_q} \succ \cdots$$

$$a_{i,j} : \beta_{S(i,1)} \succ \beta_{S(i,2)} \succ \beta_{S(i,3)} \succ \tau_1 \succ \cdots \succ \tau_m \succ$$
$$\gamma_1 \succ \cdots \succ \gamma_{3n-3m} \succ \varepsilon_1 \succ \cdots \succ \varepsilon_{n-m} \succ \boxed{\rho_{i,j}} \succ \cdots$$

$$v_t : \delta_{3m+t-1}^2 \succ \boxed{\gamma_t}$$

$$d_s^2 : \rho_{1,1} \succ \cdots \succ \rho_{n,3} \succ \boxed{\delta_s^2} \succ \cdots$$

$$w_o : \delta_{o-1}^1 \succ \boxed{\tau_o} \succ \tau_m \succ \cdots \succ \tau_1 \succ \varepsilon_{n-m} \succ \cdots \succ \varepsilon_1 \succ \cdots$$

$$w_{m+p} : \delta_{m+p-1}^1 \succ \boxed{\varepsilon_p} \succ \varepsilon_{n-m} \succ \cdots \succ \varepsilon_1 \succ \tau_1 \succ \cdots \succ \tau_m \succ \cdots$$

$$d_0^1 : \rho_{1,1} \succ \cdots \succ \rho_{n,3} \succ \nu_1 \succ \cdots \succ \nu_n \succ \boxed{\delta_0^1} \succ \cdots$$

$$d_\ell^1 : \sigma_1 \succ \cdots \succ \sigma_n \succ \chi_\ell \succ \boxed{\delta_\ell^1} \succ \cdots$$

$$d_n^1 : \sigma_1 \succ \cdots \succ \sigma_n \succ \chi_n \succ \boxed{\delta_0^2} \succ \cdots$$

$$x_i : \tau_1 \succ \cdots \succ \tau_m \succ \varepsilon_1 \succ \cdots \succ \varepsilon_{n-m} \succ \chi_i \succ \boxed{\sigma_i} \succ \cdots$$

$$y_i : \boxed{\nu_i} \succ \cdots$$

$$c_i : \boxed{\chi_i} \succ \cdots$$

with $p \in \{1, \ldots, n - m\}$, $i \in \{1, \ldots, n\}$, $q \in \{1, \ldots, 3m\}$, $t \in \{1, \ldots, 3n - 3m\}$, $s \in \{1, \ldots, 3n - 1\}$, $o \in \{1, \ldots, m\}$, and $\ell \in \{1, \ldots, n - 1\}$. Since *don't care* agents are relevant for graph-envy-freeness, we need the preferences of $d_0^1$.

Construct the directed social relationship network $G_R = (A, E_R)$ with edge set

$$
\begin{aligned}
E_R = \{&(y_i, a_{i,j}), (a_{i,j}, x_i) \mid 1 \le i \le n \text{ and } 1 \le j \le 3\} \cup \\
&\{(d_0^1, y_i) \mid 1 \le i \le n\} \cup \{(z_1, d_i^1) \mid 0 \le i \le n\} \cup \\
&\{(z_i, z_{i-1}) \mid 2 \le i \le 3m\} \cup \{(v_1, d_i^2), (d_i^2, z_{3m}) \mid 1 \le i \le 3n - 1\} \cup \\
&\{(v_i, v_{i-1}) \mid 2 \le i \le 3n - 3m\} \cup \{(x_i, c_i), (d_i^1, c_i) \mid 1 \le i \le n\} \cup \\
&\{(w_i, w_{i-1}) \mid 2 \le i \le n\} \cup \{(w_1, v_{3n-3m})\}
\end{aligned}
$$

(see Fig. 2) and the undirected social trading network $G_T = (A, E_T)$, which is constructed as a star with the *don't care* agent $d_0^1$ as its center node. Furthermore, the initial allocation $\pi_0$ is given in Table 1.

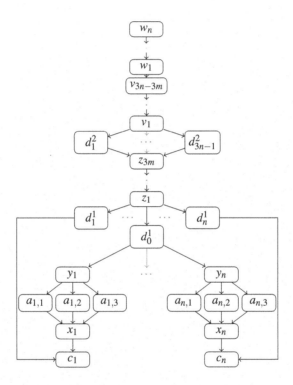

**Fig. 2.** The social relationship network $G_R = (A, E_R)$ for the proof of Theorem 4

The main idea in this construction is to introduce $n$ 3-center groups with decision maker $y_i$ and center agents $a_{i,1}, a_{i,2}, a_{i,3}$, where the agents $a_{i,j}$ are forced to receive one of the $3m$ resources resulting from the base set $B$ if their leaf $x_i$ gets an item from $R^T$. Resources from $R^T$ indicate that $i \in S$, and $R^E$ indicates $i \notin I$. Since $c_i$ will never do a trade, $x_i$ are required to get items from $R^E \cup R^T$ ($|A^X| = |R^E \cup R^T| = n$). This will help us in the second part of this proof.

**Table 1.** Initial allocation $\pi_0$ for the proof of Theorem 4

| $a$ | $\pi_0(a)$ | |
|-----|-----|-----|
| $a_{i,j}$ | $\rho_{i,j}$ | for $1 \leq i \leq n$ and $1 \leq j \leq 3$ |
| $d_i^1$ | $\delta_i^1$ | for $0 \leq i \leq n-1$ |
| $d_n^1$ | $\delta_0^2$ | |
| $d_i^2$ | $\delta_i^2$ | for $1 \leq i \leq 3n-1$ |
| $v_i$ | $\gamma_i$ | for $1 \leq i \leq 3n-3m$ |
| $w_j$ | $\tau_j$ | for $1 \leq j \leq m$ |
| $w_\ell$ | $\varepsilon_{m-\ell}$ | for $m+1 \leq \ell \leq n$ |
| $x_i$ | $\sigma_i$ | for $1 \leq i \leq n$ |
| $y_i$ | $\nu_i$ | for $1 \leq i \leq n$ |
| $z_i$ | $\beta_i$ | for $1 \leq i \leq 3m$ |
| $c_i$ | $\chi_i$ | for $1 \leq i \leq n$ |

We show that $((A, R, \succ), G_T, G_R, \pi_0, K)$ is a yes-instance of DONT-CARE-AGENT-TRADE-BASED-GEF-ALLOCATION if and only if $(B, \mathcal{S})$ is a yes-instance of X3C.

From right to left, suppose that $(B, \mathcal{S})$ is a yes-instance of X3C and let $I$ be the index set of an exact cover of $B$, i.e., a subset of $\{1, \ldots, n\}$ with $|I| = m$ such that $B = \bigcup_{i \in I} S_i$ is a disjoint union. We show that the following allocation $\pi$ is reachable from $\pi_0$:

1. $\pi(x_i) \in R^T$ if $i \in I$ and $\pi(x_i) \in R^E$ otherwise;
2. $\pi(a_{i,j}) = r_{S(i,j)}$ for $i \in I$ and $1 \leq j \leq 3$;
3. $\pi(a_{i,j}) \in R^G$ for $i \in \bar{I}$ and $1 \leq j \leq 3$;
4. every agent in $A_1^D$ gets one of her top resources;
5. every agent in $A_2^D$ gets one of her top $3n$ resources; and
6. all other agents get their most preferred resource.

Obviously, this is a graph-envy-free allocation.
Starting from $\pi_0$, perform the following trades:

$$((w_i, \pi_0(w_i)), (d_0^1, \delta_{i-1}^1)), \quad ((d_0^1, \pi_0(w_i)), (x_j, \sigma_j)), \quad ((d_0^1, \sigma_j), (d_i^1, \delta_i^1))$$

for $i = 1, \ldots, n-1$ and

$$((w_n, \pi_0(w_n)), (d_0^1, \delta_{n-1}^1)), \quad ((d_0^1, \pi_0(w_n)), (x_j, \sigma_j)), \quad ((d_0^1, \sigma_j), (d_n^1, \delta_0^2)),$$

where $\pi(x_i) \in R^T$ if and only if $i \in I$. The resulting allocation satisfies the first property of the list above.

Further, do the following steps:

$$((z_i, \beta_i), (d_0^1, \delta_{i-1}^2)), \quad ((d_0^1, \beta_i), (a_{j,\ell}, \rho_{j,\ell})), \quad ((d_0^1, \rho_{j,\ell}), (d_i^2, \delta_i^2))$$

for $i = 1, \ldots, 3n$ and

$$((v_p, \gamma_j), (d_0^1, \delta_{3m+p-1}^2)), \quad ((d_0^1, \gamma_p), (a_{j,\ell}, \rho_{j,\ell})), \quad ((d_0^1, \rho_{j,\ell}), (d_{3m+p}^2, \delta_{3m+p}^2))$$

for $p = 1, \ldots, 3n - 3m - 1$ and

$$((v_{3n-3m}, \gamma_{3n-3m}), (d_0^1, \delta_{3n-1}^2)), \quad ((d_0^1, \gamma_{3n-3m}), (a_{j,\ell}, \rho_{j,\ell}))$$

in a way that property 2 and 3 of the list above are satisfied.

After these steps, every agent in $A_2^D$ has a resource from $R^0$, every agent in $A_1^D$ (except $d_0^1$) has a resource from $R^S$, $d_0^1$ is allocated a $\rho_{i,j}$, and all other agents receive their most preferred resource, which shows that the remaining properties are satisfied.

From left to right, suppose that $(B, S)$ is a no-instance of X3C and assume that $((A, R, \succ), G_T, G_R, \pi_0, K)$ is a yes-instance of DONT-CARE-AGENT-TRADE-BASED-GEF-ALLOCATION.

Since agent $c_i$ receives $\chi_i$, every agent $x_i$ must receive resources from $R^T$ or $R^E$ to avoid graph-envy. Then every agent $a_{i,j}$ must get resources from $R^B$ if $\pi(x_i) \in R^T$ holds for their child node $x_i$. If an agent $x_i$ gets a resource from $R^E$, agent $a_{i,j}$ is forced to be allocated resources from $R^B$ or $R^G$. Due to $|R^B| = 3|R^T|$, we conclude:

1. $a_{i,j}$ must receive a resource from $R^B$ if their child node $x_i$ gets resources from $R^T$.
2. $a_{i,j}$ must receive a resource from $R^G$ if their child node $x_i$ gets resources from $R^E$.

We show that this constraint is always violated if we start with a no-instance of X3C. Due to the constructed preferences and the fact that $x_i$ has to get items from $R^T \cup R^E$, we can only distribute resources to the agents from $A^0$ after we selected (by trades) which 3-center groups represents a $S_i$ with $i \in I$. So we trade as long as $d_0^1$ receives $\delta_0^2$ and distribute the resources from $R^T \cup R^E$ to the agents from $A^X$. Now, the only agent who is allowed to trade with $d_0^1$ is $z_1$, which gives $\beta_1$ to $d_0^1$. If there is an $i \in I$ such that $b_1 \in S_i$, it can be allocated to an agent in the 3-center group of $S_i$. If this is not the case, $d_0^1$ will not find a sequence of trades such that property 1 above is satisfied. As this is a required condition, we will never find a graph-envy-free allocation. The latter case will not necessarily occur for $\beta_1$, but there is a $\beta_j$ for that $d_0^1$ cannot find a trading partner (for otherwise we would have started with a yes-instance of X3C). Thus there is no graph-envy-free allocation.                                           □

## 5  Conclusions and Open Questions

We have shown strong NP-completeness of the problems of deciding whether there is a graph-envy-free allocation in a social network such that egalitarian social welfare or the Nash product exceed a given threshold. The key idea was the construction of several 3-center groups, where every decision maker determines which resources must be allocated to her predecessors. This can be done with three-valued utility functions for GEF-EWSO, whereas GEF-NSWO needs at least four-valued utility functions. Additionally we pointed out that a missing decision possibility between two different utility values greater than 0, as appearing in binary utility functions, lead to polynomial time solvable problems. In addition, we have shown NP-hardness of the problem of whether a graph-envy-free allocation can be reached by trades in the presence of *don't care* agents. An interesting topic of future research would be to find out how the computational complexity of GEF-ESWO is affected if we restrict the problem to binary utility functions but allow $|A| < |R|$. This also applies to GEF-NSWO; moreover, it is also open how hard the problem becomes if we require, as in Theorem 1, restrictions on the utility functions. Furthermore, it would be interesting to study the complexity of more general social welfare measures like the Lorenz curve [16]. Another interesting task for future research would be to determine whether DONT-CARE-AGENT-TRADE-BASED-GEF-ALLOCATION is an NP-complete problem.[3]

# References

1. Abdulkadiroğlu, A., Sönmez, T.: Random serial dictatorship and the core from random endowments in house allocation problems. Econometrica **66**(3), 689–701 (1998)
2. Aziz, H., Bouveret, S., Caragiannis, I., Giagkousi, I., Lang, J.: Knowledge, fairness, and social constraints. In: Proceedings of the 32nd AAAI Conference on Artificial Intelligence, pp. 4638–4645. AAAI Press (2018)
3. Bentert, M., Chen, J., Froese, V., Woeginger, G.J.: Good things come to those who swap objects on paths. arXiv:1905.04219 (2019)
4. Bergson, A.: A reformulation of certain aspects of welfare economics. Q. J. Econ. **52**(2), 310–334 (1938)
5. Bouveret, S., Cechlárová, K., Elkind, E., Igarashi, A., Peters, D.: Fair division of a graph. In: Proceedings of the 26th International Joint Conference on Artificial Intelligence, pp. 135–141. AAAI Press/IJCAI (2017)
6. Bouveret, S., Chevaleyre, Y., Maudet, N.: Fair division of indivisible goods. In: Brandt, F., Conitzer, V., Endriss, U., Procaccia, A.D. (eds.) Handbook of Computational Social Choice, pp. 284–310. Cambridge University Press, Cambridge (2016)
7. Bredereck, R., Kaczmarczyk, A., Niedermeier, R.: Envy-free allocations respecting social networks. In: Proceedings of the 17th International Conference on Autonomous Agents and Multiagent Systems, pp. 283–291. IFAAMAS, July 2018
8. Chevaleyre, Y., et al.: Issues in multiagent resource allocation. Informatica **30**(1), 3–31 (2006)
9. Chevaleyre, Y., Endriss, U., Maudet, N.: Allocating goods on a graph to eliminate envy. In: Proceedings of the 22nd AAAI Conference on Artificial Intelligence, pp. 700–705. AAAI Press (2007)
10. Garey, M.R., Johnson, D.S.: Computers and Intractability: A Guide to the Theory of NP-Completeness. W. H. Freeman and Company, New York (1979)
11. Gourvès, L., Lesca, J., Wilczynski, A.: Object allocation via swaps along a social network. In: Proceedings of the 26th International Joint Conference on Artificial Intelligence, pp. 213–219. AAAI Press/IJCAI (2017)
12. Heinen, T., Nguyen, N., Nguyen, T., Rothe, J.: Approximation and complexity of the optimization and existence problems for maximin share, proportional share, and minimax share allocation of indivisible goods. J. Auton. Agents Multi-Agent Syst. **32**(6), 741–778 (2018)
13. Heinen, T., Nguyen, N.-T., Rothe, J.: Fairness and rank-weighted utilitarianism in resource allocation. In: Walsh, T. (ed.) ADT 2015. LNCS (LNAI), vol. 9346, pp. 521–536. Springer, Cham (2015). https://doi.org/10.1007/978-3-319-23114-3_31

---

[3] That the variant of this problem *without don't care agents* is in NP if the social trading network is a star can be seen as follows: Guess an allocation $\pi$ from the given instance $((A, R, \succ), G_T, G_R, \pi_0, K)$, where $G_T$ is a star, and check whether $\pi$ is reachable from $\pi_0$. Since $G_T$ in particular is a tree, this can be done in polynomial time [11, Proposition 3]. Checking whether $\pi$ is graph-envy-free can than also be done efficiently.

14. Karp, R.M.: Reducibility among combinatorial problems. In: Miller, R., Thatcher, J. (eds.) Complexity of Computer Computations, pp. 85–103. Plenum Press, New York (1972)
15. Lang, J., Rothe, J.: Fair division of indivisible goods. In: Rothe, J. (ed.) Economics and Computation. STBE, pp. 493–550. Springer, Heidelberg (2016). https://doi.org/10.1007/978-3-662-47904-9_8
16. Lorenz, M.O.: Methods of measuring the concentration of wealth. Publ. Am. Stat. Assoc. **9**(70), 209–219 (1905)
17. Papadimitriou, C.H.: Computational Complexity, 2nd edn. Addison-Wesley, Reading (1995)
18. Rothe, J.: Complexity Theory and Cryptology. An Introduction to Cryptocomplexity. EATCS Texts in Theoretical Computer Science. Springer, Heidelberg (2005). https://doi.org/10.1007/3-540-28520-2

# New Complexity Results on Aggregating Lexicographic Preference Trees Using Positional Scoring Rules

Xudong Liu[1(✉)] and Miroslaw Truszczynski[2]

[1] School of Computing, University of North Florida, Jacksonville, USA
xudong.liu@unf.edu
[2] Department of Computer Science, University of Kentucky, Lexington, USA
mirek@cs.uky.edu

**Abstract.** Aggregating *votes* that are preference orders over *candidates* or *alternatives* is a fundamental problem of decision theory and social choice. We study this problem in the setting when alternatives are described as tuples of values of attributes. The combinatorial spaces of such alternatives make explicit enumerations of alternatives from the most to the least preferred infeasible. Instead, votes may be specified implicitly in terms of some compact and intuitive preference representation mechanism. In our work, we assume that votes are given as *lexicographic preference trees* and consider two preference-aggregation problems, the *winner* problem and the *evaluation* problem. We study them under the assumption that *positional scoring rules* are used for aggregation. In particular, we consider $k$-Approval and $b$-Borda, a generalized Borda rule, and we discover new computational complexity results for them.

## 1 Introduction

Preferences are an essential component of decision making, social choice, knowledge representation, and constraint satisfaction. Fundamental problems of preference reasoning are to *aggregate* individual preference orders of a group of agents (the *votes* of agents in the group) into a consensus best candidate (the *winner*), and to identify candidates with strong consensus support from the group ("good" alternatives). These problems have been studied extensively in social choice [1]. Aggregation methods known as *positional scoring rules*, which include such well-known rules as plurality, $k$-approval and Borda, are among the best understood and the most widely used ones [4].

When the number of alternatives is small, the simplest and most effective way to describe a preference order (a vote) is to enumerate the alternatives from the most to the least preferred. Moreover, given a collection of such votes, for many aggregation rules, including all that are based on positional scoring, computing winners and "good" candidates is easy — it can be done in polynomial time. The situation changes when alternatives are characterized in terms of *attributes*

© Springer Nature Switzerland AG 2019
S. Pekeč and K. B. Venable (Eds.): ADT 2019, LNAI 11834, pp. 97–111, 2019.
https://doi.org/10.1007/978-3-030-31489-7_7

(or issues), and are specified by tuples of attribute values. Spaces of such alternatives, often called *combinatorial domains*, are large. Indeed, the number of alternatives grows exponentially with the number of attributes. This large size of combinatorial domains brings up two problems. First, it is no longer feasible to describe votes by enumerating alternatives in the order of preference. Thus, formalisms offering compact and intuitive representations of votes are needed. Several such *preference formalisms* have been developed over the years including penalty logic [5], possibilistic logic [6], conditional preference networks (CP-nets) [3], and variations of lexicographic preference trees and forests [2,12–14]. Second, when votes are given as expressions in some preference formalism, computing the winner or a "good" candidate is no longer easy. In fact, it is known that for many preference formalisms these problems are NP-hard even when positional scoring rules are used to aggregate votes. *Issue-by-issue* aggregation addresses the computational hardness problem but often leads to results different from those obtained by applying common voting rules [7].

In this paper, we assume that votes are represented as *lexicographic preference trees*, or *LP-trees*, for short [2], and that they are aggregated by some simple positional scoring rule such as $k$-Approval and $b$-Borda, a generalized Borda rule. Given this setting, we study the problem of computing the best alternative, and its related problem to decide whether an alternative with the score exceeding a given threshold (a "good" alternative) exists. We refer to the former problem as the *winner* problem and to the latter one as the *evaluation* problem. LP-trees are the votes to aggregate. Depending on the structure of the trees and whether local preferences at any node rely on evaluations of ancestor nodes, LP-trees are classified into *unconditional importance and unconditional preference* (UI-UP) trees, *unconditional importance and conditional preference* (UI-CP) trees, *conditional importance and unconditional preference* (CI-UP) trees, and *conditional importance and conditional preference* (CI-CP) trees.

These problems have been studied in the literature and turn out often computationally hard. For $k$-approval, for some specific values of $k$ (e.g., exactly half of the number of the alternatives), both problems are in P across all four classes; however, for some other, they are NP-hard and NP-complete, respectively, for all classes of trees [10]. For the standard Borda rule, the *winner* problem and the *evaluation* problem are in P for UI-UP trees, but are NP-hard and NP-complete, for all other classes [10]. For all four classes of LP-trees, deciding if a given outcome is a Condorcet winner is coNP-hard [10]. Liu and Truszczynski [11] showed, for $(k, l)$-Approval, a two level approval rule, that the two problems are NP-hard and NP-complete for all four classes, when $k$ and $l$ are of some specific values.

Our main contributions are algorithms and complexity results for the *winner* and the *evaluation* problems when votes are specified as LP-trees aggregated by two positional scoring rules: $k$-approval and $b$-Borda. Specifically, for the $k$-approval rule, we propose polynomial time algorithms to solve these problems for all classes of LP-trees, when $k$ intuitively is within a polynomial difference from half of the number of alternatives. For the $b$-Borda rule, a generalized Borda rule, we prove NP-hardness of the two problems for all classes of LP-trees for every fixed positive integer $b$.

## 2    Technical Preliminaries

A *vote* over a set $\mathcal{X}$ of *alternatives* (or *outcomes*) is a *strict total* order $\succ$ on $\mathcal{X}$. In this work, we consider votes over alternatives from *combinatorial domains* determined by a set $\mathcal{A}$ of $p$ binary *attributes* $X_1, X_2, \ldots, X_p$, with each attribute $X_i$ having a binary domain $D(X_i) = \{0_i, 1_i\}$. The combinatorial domain in question is then the set $\mathcal{X}(\mathcal{A}) = D(X_1) \times D(X_2) \times \ldots \times D(X_p)$. If $\mathcal{A}$ is implied by the context, we write $\mathcal{X}$ instead of $\mathcal{X}(\mathcal{A})$. Let $o \in \mathcal{X}$ be an alternative. We will use $o(X_i)$ to denote the value of attribute $X_i$ in $o$, and $o|_S$ the partial alternative projected from $o$ to $S \subseteq \mathcal{A}$.

Clearly, the cardinality of $\mathcal{X}(\mathcal{A})$ is $2^p$. Thus, even for relatively small values of $p$, eliciting precise orders over all alternatives and representing them directly may be infeasible. Instead, in cases when votes have some structure, they often can be represented compactly by means of intuitive "preference expressions" in logical or graphical formalisms [3,8,9].

In this work we focus on one such formalism, *lexicographic preference trees* or *LP-trees*, for short [2]. An *LP-tree* over a set $\mathcal{A}$ of $p$ binary attributes $X_1, \ldots, X_p$ is a *binary tree*. Each non-leaf node $t$ is labeled by an attribute from $\mathcal{A}$, denoted by $Iss(t)$. Every non-leaf node $t$ has two outgoing edges, each labeled by a distinct value in $D(Iss(t))$. The two outgoing edges represent the preference information over $D(Iss(t))$: the value labeling the left edge is preferred to the value labeling the right. Leaf nodes are empty boxes. In addition, we require that each attribute appears *exactly once* on each path from the root to a leaf.

Intuitively, the attribute labeling the root of an LP-tree is of highest importance. Alternatives with the more preferred value of that attribute are preferred over alternatives with the other (less preferred) value. The two subtrees further refine that ordering. The left subtree determines the ranking of alternatives in the more preferred "upper half" and the right subtree determines the ranking of alternatives in the less preferred "lower half." In each case, the same principle is used recursively, with the attribute labeling the root of the subtree being the most important one. The attributes labeling the roots of the subtrees need not be the same (the relative importance of attributes may depend on values for their "ancestor" attributes labeling the nodes on the path to the root).

Given an alternative $o = x_1 x_2 \ldots x_p$, we find its preference rank in $T$ by traversing the LP-tree from the root to a leaf. When at node $t$ labeled with the attribute $X_i$, we follow down to the left subtree if $x_i$ is preferred. Otherwise, we follow down to the right subtree. In this way, we end up in a unique leaf of the tree. Let us denote its label by $r(o)$. We take $r(o)$ as the *preference rank* of $o$ and say that an alternative $o$ is (strictly) *preferred* (in $T$) to an alternative $o'$ precisely when $r(o) < r(o')$ (informally, alternatives with lower ranks are preferred to those that are ranked higher). In this way, $T$ defines a vote (a total order).

To illustrate these concepts, let us consider an example setting with three binary attributes describing meal alternatives. The *Appetizer* can be either *salad* (*s*) or *soup* (*u*), the *Entree* could be either *beef* (*b*) or *fish* (*f*), and the *Drink* could be *beer* (*r*) or *wine* (*w*). Looking at the LP-tree in Fig. 1a, we see that the most important attribute is Entree (at the root) with fish preferred over beef. Among meal alternatives with fish entree, the most important attribute is Drink, for which wine is preferred to beer. Same preference among meals with beef entree. Similar reasoning applies to the rest of the tree. Clearly, the most preferred meal alternative is salad, fish and wine.

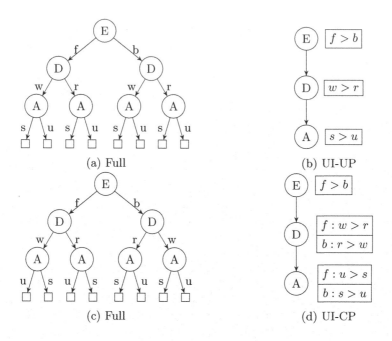

**Fig. 1.** Unconditional importance LP-trees

A full representation of an LP-tree requires as much space as an explicit enumeration of the preference order. However, sometimes LP trees can be represented in a much more concise way. For instance, if for some node $t$, its two subtrees are identical (that is, the corresponding nodes are assigned the same attribute), they can be collapsed to a single subtree, with the same assignment of attributes to nodes. This subtree is now a single child of its parent and it is neither left nor right, as it is associated with *each* of the two values of its parent node attribute. To retain the preference information, at each node $t'$ of the subtree we place a *conditional preference table*, and each preference in it specifies the preferred value for the attribute labeling $t'$ given the value of the attribute labeling $t$. In the extreme case when for every node its two subtrees are identical, the tree can be collapsed to a path.

We now formally extend the definition of an LP-tree to cover the case with collapsed subtrees. From now on, an *LP-tree* is a tree in which every node except for the leaves has either two children, the left child and the right child, or a single "straight down" child, representing the collapse of the subtrees. The nodes are labeled with attributes and, as before, we assume that each attribute appears exactly once on each path from the root to a leaf. Further, each node in an LP-tree is assigned a *conditional preference table* (CPT) compensating for the loss in the structure due to the collapse higher in the tree. To specify these conditional preferences, we define $Par(t) \subseteq one(t)$ to be the set of *parent* nodes of $t$, where $one(t)$ denotes the set of ancestor nodes of $t$ with just one child.

We can now introduce a useful classification of LP-trees. If for every node $t$ in an LP-tree, $Par(t) = \emptyset$, all (local) preferences are unconditional and all CPTs consist of a single entry. Such trees are called *unconditional preference* LP-trees (or UP trees, for short). Similarly, LP-trees with all non-leaf nodes having exactly one child, are called an *unconditional importance* LP-trees (UI trees, for short). Combining these characteristics leads to four classes of *collapsed* LP-trees: unconditional importance and unconditional preference LP-trees (UI-UP trees), unconditional importance and conditional preference trees (UI-CP trees), conditional importance and unconditional preference trees (CI-UP trees), and conditional importance and conditional preference trees (CI-CP trees). CI-CP trees are most expressive, capturing all LP-trees, UI-UP trees are the least expressive of the four collapsed types.

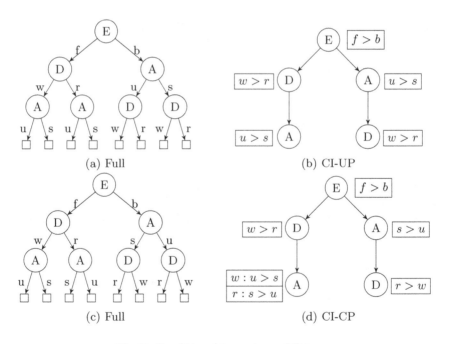

**Fig. 2.** Conditional importance LP-trees

We present examples of UI trees in Fig. 1 and CI trees in Fig. 2.

For an LP-tree $T$ of any class, we denote by $|T|$ the size of $T$, that is, the total size of CPT's associated with the nodes in $T$. The size of a CPT of a node is the total size of rows in it, each of which is measured as the number of values in the condition plus 1 for the preferred value in the domain of the attribute labeling the node. Particularly, the size of an UI-UP LP-tree is the number of nodes in it.

Finally, we recall the concept of a profile and of a positional scoring rule. A set of votes over a domain $\mathcal{X}$ (collected from, say, $n$ voters) is called a *profile*. The size of a profile is the total size of the votes inside it. Among many rules proposed to aggregate a profile into a single preference ranking representing the group, positional scoring rules have received particular attention. For profiles over a domain with $h$ alternatives, a *scoring vector* is a sequence $w = (w_0, \ldots, w_{h-1})$ of integers such that $w_0 \geq w_1 \geq \ldots \geq w_{h-1}$ and $w_0 > w_{h-1}$. Given a vote $v$ with the alternative $o$ in position $i$ $(0 \leq i \leq h - 1)$, the score of $o$ in $v$ is given by $s_w(o, v) = w_i$. Given a profile $P$ of votes and an alternative $o$, the score of $o$ in $P$ is given by

$$s_w(o, P) = \sum_{v \in P} s_w(o, v).$$

These scores determine the ranking generated from $P$ by the scoring vector $w$ (assuming, as is common, some independent tie breaking rule). In this paper we consider two positional scoring rules:

1. $k$-approval: $(1, \ldots 1, 0, \ldots 0)$ with $k$, $1 \leq k \leq 2^p$ the number of 1's
2. $b$-Borda: $(2^{p-b} - 1, 2^{p-b} - 2, \ldots, 1, 0, \ldots, 0)$ with $0 \leq b < p$

We see that the traditional Borda rule is 0-Borda, and that (p-1)-Borda is the Plurality rule.

## 3   Computing Ranks

We now show how to compute the *rank* of an outcome in an LP-tree (possibly in the collapsed form). For our algorithms and complexity results (membership proofs) it is important that we can compute ranks (and so, also scores) of outcomes in polynomial time.

Given an LP-tree $T$ over $\mathcal{A} = \{X_1, \ldots, X_p\}$ and an outcome $o \in \mathcal{X}(\mathcal{A})$, the computation of the rank $r(o, T)$ of $o$ in $T$ is given below, where we write $rt(T)$ for the root of a tree $T$.

First, we set $r = 0$ and $T' = T$ that represent the rank we are computing and the subtree under consideration, respectively. Second, for $i$ from 1 to $p$, we repeatedly perform the next three steps: (1). Set $Y = iss(rt(T'))$, and identify the preference rule in the CPT of form $o|iss(Par(rt(T'))) : y > y'$; (2). If $o(Y) \neq y$, update $r = r + 2^{p-i}$; (3). If $rt(T')$ has a single subtree $T'_s$, update $T' = T'_s$; otherwise, $rt(T')$ has two subtrees $T'_l$ (left subtree, preferred) and $T'_r$ (right subtree, less preferred), then we update $T' = T'_l$ if $o(Y) = y$, or $T' = T'_r$ if $o(Y) = y'$.

Clearly, this algorithm applies to any of the four classes of LP-trees, and the running time of this algorithm is $O(p)$, for it walks down the LP-tree from the root to a leaf and the height of the tree is bounded by $O(p)$.

Now that ranks are computed, calculating the scores of an outcome for $k$-approval and $b$-Borda is now straightforward. Specifically,

1. $k$-approval: $s_{kApp}(o, T) = 1$, if $r(o, T) < k$; 0, otherwise.
2. $b$-Borda: $s_{bB}(o, T) = 2^{p-b} - 1 - r(o, T)$, if $r(o, T) < 2^{p-b}$; 0, otherwise.

For example, the rank of meal alternative with soup, fish and beer in the UI-CP tree in Fig. 1d is 2, the summation of 0 (fish is the more preferred), $2^{3-2}$ (beer is the less preferred given fish), and 0 (soup is the more preferred given fish). In the full representation in Fig. 1c, we clearly see our alternative reaches the third leaf counted from the left-most one. Therefore, its 4-Approval score and its 1-Borda score are both 1.

## 4    The Problems and Their Complexity

From now on, we only consider positional scoring rules without the scoring vectors explicitly given, rather defined by a polynomial time algorithm that returns the score of an outcome based on the rank of it and the total number of outcomes. Clearly, Both $k$-Approval and $b$-Borda fit in this category. This is so because the scoring vectors for combinatorial domains on which LP-trees are specified are of size exponential in $p$, making our aggregation problems solvable in time polynomial in this sheer size.

Let us fix such a positional scoring rule $\mathcal{D}$. Given a profile $P$, the *winner* problem under $\mathcal{D}$ is to compute an alternative $o \in \mathcal{X}$ with the maximum score $s_{\mathcal{D}}(o, P)$. Similarly, given a profile $P$ and a positive integer $R$, the *evaluation* problem for $\mathcal{D}$ asks if there exists an alternative $o \in \mathcal{X}$ such that $s_{\mathcal{D}}(o, P) \geq R$. In each case, $s_{\mathcal{D}}(o, P) = \sum_{T \in P} s_{\mathcal{D}}(o, T)$.

We apply the voting rules listed above to profiles consisting of LP-trees. We distinguish four classes of profiles, UI-UP, UI-CP, CI-UP and CI-CP, depending on the type of LP-trees they consist of.

### 4.1    $k$-Approval

If $k = 2^{p-1}$ the evaluation problem is in P for all four classes of profiles of LP-trees [9]. However, the problem is NP-complete, again for all four types of profiles, when $k = \alpha \cdot 2^p$ where $\alpha$ is a rational number of form $a/2^p$ for any integer $1 \leq a < 2^p$ and $\alpha \neq 1/2$ [10]. Clearly, in each case where the evaluation problem is NP-complete, the winner problem is NP-hard.

Extending the above polynomial time result for $2^{p-1}$-Approval, we show that, for all of the four classes {UI,CI}-{UP,CP} of profiles, the two problems stay in P for all values $k$ that differ from $2^{p-1}$ by a polynomial in $p$. In other words, for every $k \in [2^{p-1} - f(p), 2^{p-1} + f(p)]$, where $f(p)$ is a polynomial in $p$ such that $0 < f(p) < 2^{p-1}$, both the winner and the evaluation problems for $k$-approval can be solved by polynomial time algorithms. The next two results address the two cases for $k$, respectively.

**Theorem 1.** *Let $f(p)$ be a polynomial in $p$ such that $0 < f(p) < 2^{p-1}$ for all $p \geq 1$. The winner problem under $k$-approval, where $k = 2^{p-1} + f(p)$, for any profile of LP-trees of any class in $\{UI, CI\}$-$\{UP, CP\}$, can be solved in time polynomial in the size of the profile.*

*Proof.* For an outcome $o$, we write $s(o)$ and $s'(o)$ for the $k$-approval and $2^{p-1}$-approval scores of $o$ in the profile $P$ of any of the four classes. Since $k \geq 2^{p-1}$, $s(o) \geq s'(o)$ holds for every outcome $o$.

Next, we write $P_i$ for the set of votes (LP-trees) in $P$ whose roots are labeled with $X_i$ ($i = 1, \ldots, p$). We define $b_{i,0}$ to be the number of votes in $P_i$ with 0 preferred to 1 at the root and, similarly, $b_{i,1}$ to be the number of votes in $P_i$ with 1 preferred to 0 at the root. Clearly, for an outcome $o = \langle o_1, \ldots o_p \rangle$, we have

$$s'(o) = \sum_{i=1}^{p} \epsilon_i(o),$$

where $\epsilon_i(o) = b_{i,0}$ if $o_i = 0$, and $\epsilon_i(o) = b_{i,1}$ if $o_i = 1$.

Let us define $x_i = 0$, if $b_{i,0} > b_{i,1}$, $x_i = 1$, if $b_{i,1} > b_{i,0}$, and $x_i = *$, otherwise. (Symbol '$*$' can be either 0 or 1.) In the first case, picking 0 for position $i$ maximizes the contribution of the votes in $P_i$ to the $2^{p-1}$-approval score (and picking 1 is strictly worse). In the second case, the situation is dual. In the third case, the choice of the value for position $i$ does not affect the $2^{p-1}$-approval score. Thus, the tuple $\langle x_1, \ldots x_p \rangle$ is a description of all those tuples that have the largest $2^{p-1}$-approval score votes in the profile $P$; they are those tuples that can be obtained from $\langle x_1, \ldots x_p \rangle$ by arbitrarily instantiating $*$'s with 0's and 1's.

Let $S$ be the set of all outcomes $o$ such that $s(o) > s'(o)$. Clearly, every outcome $o \in S$ must be ranked between $2^{p-1}$ and $2^{p-1} + f(p)$ in some tree of $P$. Thus, we can construct $S$ by taking the union $\bigcup_{T \in P} \{o \in \mathcal{X} : 2^{p-1} < r(o, T) \leq 2^{p-1} + f(p)\}$. We see that $|S| \leq nf(p)$. Since $p$ and $n$ are bounded by $size(P)$, $|S| = O(h(size(P)))$, for some polynomial $h$. Hence, the construction of $S$ takes time polynomial in $size(P)$. Importantly, for every outcome $o$ not in $S$, $s(o) = s'(o)$.

Let $o'$ be an outcome in $S$ with the maximum $k$-approval score ($o'$ can be computed in polynomial time in $size(P)$), and let $o''$ be any outcome obtained by instantiating all $*$'s in $\langle x_1, \ldots, x_p \rangle$ to 0's and 1's (also can be computed in time $O(size(P))$). Then, there are the following two cases for their $k$-approval scores $s(o')$ and $s(o'')$.

(1). If $s(o') \geq s(o'')$, then for every outcome $o \notin S$, $s(o') \geq s(o'') \geq s'(o'') \geq s'(o) = s(o)$, and (by the choice of $o'$), for every $o \in S$, $s(o') \geq s(o)$. Thus, $o'$ has the largest $k$-approval score among all outcomes.

(2). If $s(o'') > s(o')$, then for every $o \in S$, $s(o'') > s(o)$ (by the choice of $o'$), and for every $o \notin S$, $s(o'') = s'(o'') \geq s'(o) = s(o)$. Thus, in this case, $o''$ has the largest $k$-approval score among all outcomes.

Therefore, to compute the winner under $k$-approval for $2^{p-1} \leq k \leq 2^{p-1} + f(p)$, all we need is to compute the above-mentioned $o'$ and $o''$ in polynomial time and return one of them with the bigger $k$-approval score. □

**Theorem 2.** *Let $f(p)$ be a polynomial in $p$ such that $0 < f(p) < 2^{p-1}$ for all $p \geq 1$. The winner problem under $k$-approval, where $k = 2^{p-1} - f(p)$, for any profile of LP-trees of any class in $\{UI, CI\}$-$\{UP, CP\}$, can be solved in time polynomial in the size of the profile.*

*Proof.* We will use $s'$ and $s$ in the same sense as above. However, since now $k \leq 2^{p-1}$, $s'(o) \geq s(o)$ holds for every outcome $o$. We define $A$ to be the set of all outcomes $o$ such that $s'(o) > s(o)$. Clearly, for every outcome $o$ not in $A$, $s'(o) = s(o)$. As before, the set $A$ can be computed in time bounded by a polynomial in $size(P)$. If $A$ contains every outcome (that is, $|A| = 2^p$), then we compute an outcome with the highest $k$-approval score by computing the $k$-approval scores of all outcomes in $A$ and selecting the one with the highest score. Since the size of $A$ is polynomial in the size of the profile, the task takes polynomial time (in the size of the profile).

The case when $|A| < 2^p$ is harder. To address it, let us assume that we have computed the set $B$ of top $|A|+1$ outcomes according to their $s'$-score (the $2^{p-1}$-approval score), as well as an outcome $o$ in $B \cup A$ with the maximum $k$-approval score ($s$-score).

We claim that $o$ is an outcome with the maximum $k$-approval score over all outcomes. Indeed, consider an arbitrary outcome $o'$. If $o' \in B \cup A$, then $s(o) \geq s(o')$ (by the way $o$ was selected). Thus, let us assume that $o' \notin B \cup A$. Since $|B| > |A|$, there is at least one outcome $o'' \in B \setminus A$. Because $o'' \in B$, $s(o) \geq s(o'')$. Moreover, since $o'' \notin A$, $s(o'') = s'(o'')$ and, since $o'' \in B$ and $o' \notin B$, $s'(o'') \geq s'(o')$. Finally, since $o' \notin A$, $s'(o') = s(o')$. Combining these four inequalities, we obtain that $s(o) \geq s(o')$. Thus, the claim follows.

Clearly, $|B \cup A| \leq 2|A|+1$ is bounded by a polynomial in $size(P)$. Thus, once $B$ is computed, finding an alternative in $B \cup A$ with the highest $k$-approval score can be done in time polynomial in $size(P)$. To complete the proof, it suffices then to show how to compute $B$ in polynomial time.

To this end, for each $i = 1, \ldots, p$, we set $d_i = |b_{i,0} - b_{i,1}|$ ($b_{i,0}$ and $b_{i,1}$ are as in the previous proof). We also select any outcome that has the highest $s'$-score (we explained in the previous proof how to compute it in polynomial time) and denote it by $o$. Finally, we compute the score $s'(o)$.

Let $S \subseteq \{1, \ldots, p\}$ be a set of attribute indices, and let $o_S$ be an alternative obtained from $o$ by "flipping" its values in positions in $S$. Every alternative can be described in these terms. This is useful as the $s'$-score of $o_S$ is easy to compute. Namely, we have

$$s'(o_S) = s'(o) - w(S),$$

where $w(S) = \sum_{i \in S} d_i$ is the *weight* of $S$.

It follows that $B$ consists of $|A|+1$ smallest-weight subsets of $\{1, \ldots, p\}$. We will now show that given a list $D = \{d_1, d_2, \ldots, d_p\}$ ($d_i \geq 0$, for $i = 1, \ldots, p$) and an integer $t$, the $t+1$ smallest-weight subsets of $\{1, \ldots, p\}$ can be computed in time bounded by a polynomial in $p$ and $t$.

Let $r$ be an integer such that $2^r \geq t+1$. Let us assume that $L_r$ is the set of $t+1$ smallest-weight subsets of $\{1,\ldots,r\}$. Let

$$L'_{r+1} = L_r \cup \{S \cup \{r+1\} \colon S \in L_r\}$$

and let $L_{r+1}$ be the collection of $t+1$ smallest-weight subsets $S$ of $L'_{r+1}$. We will show that $L_{r+1}$ contains $t+1$ smallest-weight subsets $S$ of $\{1,\ldots,r+1\}$. Indeed, let us consider $S \subseteq \{1,\ldots,r+1\}$ such that $S \notin L'_{r+1}$. If $S \subseteq \{1,\ldots,r\}$, then $S \notin L_r$. Thus, $w(S) \geq w(S')$, for every $S' \in L_r$. If $r+1 \in S$, then $S = R \cup \{r+1\}$, for some $R \subseteq \{1,\ldots,r\}$. Since $S \notin L'_{r+1}$, $R \notin L_r$. Thus, $w(R) \geq w(R')$, for every $R' \in L_r$ and so, $w(S) = w(R \cup \{r+1\}) \geq w(R' \cup \{r+1\})$ for all $R' \in L_r$. In each case, it follows that there are at least $t+1$ sets $S'$ in $L'_{r+1}$ such that $w(S) \geq w(S')$. Thus for every $S' \in L_{r+1}$, $w(S) \geq w(S')$.

Clearly, the list $L_p$ consists of $|A|+1$ smallest weight subsets of $\{1,\ldots,p\}$. Thus, it can be taken for $B$. To compute it, we first find the smallest $r$ such that $2^r \geq |A|+1$ (such an $r$ exists as we are now considering the case when $|A| < 2^p$). We then construct the collection $U$ of all subsets of $\{1,\ldots,r\}$ (this collection has no more than $2|A|$ elements and can be constructed in time bounded by a polynomial in $p$ and $|A|$). Next, we construct $L_r$ by selecting from $U$ its $|A|+1$ smallest-weight elements. Since $|U| \leq 2|A|$, this task also can be accomplished in polynomial time (in $p$ and $|A|$).

From now on, we construct $L_{r+1}, L_{r+2}, \ldots L_p$ recursively, as described above. Since each step of the construction can be accomplished by the same polynomial-time algorithm (form the collection $L'$, select its $|A|+1$ smallest-weight elements to form the next $L$), and since the number of steps is bounded by $p$, the total time needed to construct $B$ ($L_p$) is bounded by a polynomial in $p$ and $|A|$.    □

## 4.2  $b$-Borda

The rule $b$-*Borda* is a generalized Borda rule with the scoring vector $(2^{p-b} - 1, 2^{p-b} - 2, \ldots, 1, 0, \ldots, 0)$. If $b = 0$, $b$-Borda, or 0-Borda, is reduced to the standard Borda rule. For the most restrictive case of UI-UP profiles, the evaluation and winner problems for 0-Borda rule are in P; for the other three classes of profiles, they are NP-complete and NP-hard respectively [9,10].

However, when $b > 0$, we show that for every fixed value of $b$, the winner and the evaluation problems under $b$-Borda are NP-hard and NP-complete, respectively, no matter what the type of LP-trees used in the profiles. The cases of UI-CP, CI-UP and CI-CP trees are handled by a fairly direct reduction from the corresponding problems under the standard Borda rule. The case of UI-UP profiles requires a different argument (the winner and the evaluation problems under the standard Borda rule are, as we noted, in P). We start with the latter.

**Theorem 3.** *The evaluation problem under 1-Borda for the class of UI-UP profiles over $p > 1$ binary attributes is NP-complete.*

*Proof.* We show that the evaluation problem is NP-complete. The membership in NP is obvious. The NP-hardness follows from a polynomial reduction from the 2-MINSAT[1] problem.

Given a 2-MINSAT instance $(\Phi, l)$, where $\Phi$ consists of 2-clauses $C_1, \ldots, C_m$ over variables $I = \{X_1, \ldots, X_p\}$, we construct an instance of our problem as follows.

First, we introduce a new binary variable $X_q$ and define the set of attributes $\mathcal{A}$ by setting $\mathcal{A} = \{X_1, \ldots, X_p, X_q\}$.

Second, for each $C_i \in \Phi$, we now build a set $P_i$ of 12 UI-UP LP-trees over $\mathcal{A}$. As an example, w.l.o.g, let $C_i$ be $\neg X_2 \vee X_4$[2]. The fragment of the profile determined by $C_i$ is given by the multi-set

$$P_i = \{B_{i_1}, B_{i_2}, B_{i_1}, B_{i_2}, B_{i_1}, B_{i_2}, B'_{i_1}, B'_{i_2}, B''_{i_1}, B''_{i_2}, B''_{i_1}, B''_{i_2}\},$$

where the trees $B_{i_1}, B_{i_2}, B'_{i_1}, B'_{i_2}, B''_{i_1}$, and $B''_{i_2}$ are shown in Fig. 3. (Note that in each of these trees the dotted edge represents the subtree from $X_2$ to $X_{p-1}$ where preferences are unanimous and given as the label right next to the dotted edge.) In other words, the profile $P_i$ contains three copies of $B_{i_1}$ and $B_{i_2}$, one copy of $B'_{i_1}$ and $B'_{i_2}$, and two copies of $B''_{i_1}$ and $B''_{i_2}$. We define the overall profile $P$ as the collection of all profiles $P_i$, $1 \leq i \leq m$. That is, $P = \bigcup_{1 \leq i \leq m} P_i$. Clearly, we have $12 \cdot m$ UI-UP LP-trees in the profile $P$.

Finally, we set the threshold value $R = 15a \cdot (m - l) + 3a \cdot l$, where we use $a$ to denote $2^{p-1}$.

Let $o$ be an outcome over $\mathcal{A}$. Let $B$ be a UIUP tree over $\mathcal{A}$, $X_j$ the most important attribute of $B$. We define the 1-Borda score of $o$ in tree $B$, denoted by $s_{1B}(o, B)$, to be 0 if outcome $o$ has the non-preferred value on $X_j$; $s_{Borda}(B_{|\mathcal{A} \setminus \{X_j\}}, o_{|\mathcal{A} \setminus \{X_j\}})$, otherwise. We now compute the 1-Borda score of $o$ according to whether it satisfies $X_q$ and $C_i$. If $o \models X_q \wedge \neg C_i$, that is, $o \models X_q \wedge X_2 \wedge \neg X_4$, we have

$$s_{1B}(o, P_i) = \underbrace{(2^p - 1 + 2^{p-1} + 1) * 3}_{\text{three copies of } B_{i_1} \text{ and } B_{i_2}} + \underbrace{(0)}_{B'_{i_1} \text{ and } B'_{i_2}} + \underbrace{(2^p - 1 + 2^{p-1} + 1) * 2}_{\text{two copies of } B''_{i_1} \text{ and } B''_{i_2}}$$

$$= 15a.$$

If $o \models X_q \wedge C_i$, we need to consider three cases:
(1). If $o \models X_q \wedge \neg X_2 \wedge X_4$, we have

$$s_{1B}(o, P_i) = \underbrace{(0) * 3}_{\text{three copies of } B_{i_1} \text{ and } B_{i_2}} + \underbrace{(2^p - 1 + 2^{p-1} + 1)}_{B'_{i_1} \text{ and } B'_{i_2}} + \underbrace{(0) * 2}_{\text{two copies of } B''_{i_1} \text{ and } B''_{i_2}}$$

$$= 3a.$$

---

[1]  Given a set $\Phi$ of $n$ 2-clauses $\{C_1, \ldots, C_n\}$ over a set of propositional variables $\{X_1, \ldots, X_p\}$, and a positive integer $l$ ($l \leq n$), decide whether there is a truth assignment that satisfies at most $l$ clauses in $\Phi$.

[2]  We will build $P_i$ according to what $C_i$ contains: the two atoms in $C_i$ are the labels of the top two levels of trees, and whether the atom is negated affects the preference on that atom.

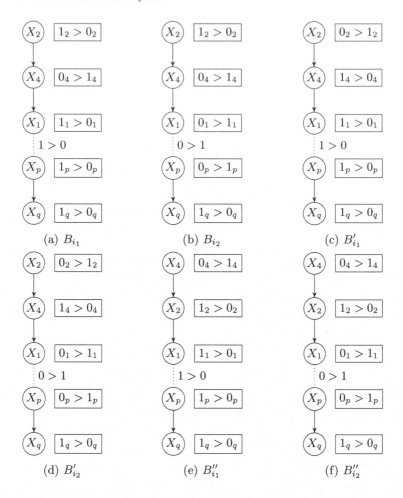

**Fig. 3.** Construction UI-UP trees in the proof of Theorem 3

(2). If $o \models X_q \wedge \neg X_2 \wedge \neg X_4$, we have

$$s_{1B}(o, P_i) = \underbrace{(0) * 3}_{\text{three copies of } B_{i_1} \text{ and } B_{i_2}} + \underbrace{(2^{p-1} - 1 + 1)}_{B'_{i_1} \text{ and } B'_{i_2}} + \underbrace{(2^{p-1} - 1 + 1) * 2}_{\text{two copies of } B''_{i_1} \text{ and } B''_{i_2}}$$

$$= 3a.$$

(3). If $o \models X_q \wedge X_2 \wedge X_4$, we have

$$s_{1B}(o, P_i) = \underbrace{(2^{p-1} - 1 + 1) * 3}_{\text{three copies of } B_{i_1} \text{ and } B_{i_2}} + \underbrace{(0)}_{B'_{i_1} \text{ and } B'_{i_2}} + \underbrace{(0)}_{\text{two copies of } B''_{i_1} \text{ and } B''_{i_2}}$$

$$= 3a.$$

Thus, for $o \models X_q \wedge C_i$, we have $s_{1B}(o, P_i) = 3a$.

Similarly, we can compute that $s_{1B}(o, P_i) < 15a$, if $o \models \neg X_q \wedge \neg C_i$; and $s_{1B}(o, P_i) < 3a$, if $o \models \neg X_q \wedge C_i$.

We now show that there exists an outcome over $\mathcal{A}$ with 1-Borda score at least $R$ if and only if there exists an assignment over $I$ that satisfies at most $l$ clauses in $\Phi$.

($\Leftarrow$) We assume there is an assignment $v$ over $I$ satisfying at most $l$ clauses in $\Phi$. Define an outcome $o = (v, 1_q)$. It is clear that $s_{1B}(o, P) \geq R$.

($\Rightarrow$) We assume there is an outcome $o$ over $\mathcal{A}$ such that $s_{1B}(o, P) \geq R$. If $o \models \neg X_q$, we could flip the value on $X_q$ from $0_q$ to $1_q$, and obtain $o'$ such that $s_{1B}(o', P) > s_{1B}(o, P) \geq R$. Assuming $o'|_I$ satisfies $l'$ ($l' > l$) clauses in $\Phi$, we have that $s_{1B}(o', P) = 15a \cdot (m - l') + 3a \cdot l' > R$; thus, $l' < l$. A contradiction! Otherwise, if $o \models X_q$, we are done.    □

**Theorem 4.** *Let $b$ be an arbitrary integer such that $b > 1$. The evaluation problem under $b$-Borda for the class of UI-UP profiles over $p > b$ binary attributes is NP-complete.*

*Proof.* Clearly, the problem is in NP. To show it is NP-hard, we reduce to this problem the NP-complete problem we proved in Theorem 3, for which we call $1\text{-}Borda_{UI\text{-}UP}^{ev}$.

Given an instance $\langle \mathcal{A}, P, l \rangle$ of $1\text{-}Borda_{UI\text{-}UP}^{ev}$, where $\mathcal{A} = \{X_1, \ldots, X_p\}$ is a set of $p$ attributes, $P = \langle T_1, \ldots, T_m \rangle$ is a profile of $m$ UI-UP trees over $\mathcal{A}$, and $l$ is a positive integer, we construct an instance $\langle \mathcal{A}', P', l' \rangle$ of our problem as follows.

We first build $\mathcal{A}' = \{X_1, \ldots, X_p, Y_1, \ldots, Y_{b-1}\}$ with $b - 1$ new attributes $Y_i$. Then, for each $T_i \in P$, we build $T_i'$ to be $Y_1 \triangleright \ldots Y_{b-1} \triangleright T_i$, and we get $P' = \{T_1', \ldots, T_m'\}$. Lastly, we set $l' = l$.

Let $o$ be an arbitrary outcome over $\mathcal{A}$. We consider outcome $o'$ over $\mathcal{A}'$ such that $o'|_{\mathcal{A}} = o$ and $o'(Y_i) = 1$ for all $Y_i$. For any tree $T_i \in P$, we let $r(o, T_i)$ and $r(o', T_i')$ be the ranks of $o$ and $o'$ in $T_i$ and $T_i'$, respectively. Let $X_{i_1}$ be the root attribute in $T_i$. The 1-Borda score $s_{1B}(o, T_i) = 2^{p-1} - 1 - r(o, T_i)$, if $o(X_{i_1})$ is the preferred value of $X_{i_1}$; $s_{1B}(o, T) = 0$, otherwise. We also have the $b$-Borda score $s_{bB}(o', T_i') = 2^{p+b-1-b} - 1 - r(o', T_i') = 2^{p-1} - 1 - r(o', T_i')$; $s_{bB}(o', T_i') = 0$, otherwise. Because we know $r(o, T_i) = r(o', T_i')$, we derive that $s_{1B}(o, T_i) = s_{bB}(o', T_i')$.

It is simple to see that there is an outcome with 1-Borda score at least $l$ for profile $P$ if and only if there is an outcome with $b$-Borda score at least $l'$ for profile $P'$.    □

**Theorem 5.** *Let $b$ be an arbitrary integer such that $b \geq 1$. The evaluation problem under $b$-Borda for the class of CI-UP (UI-CP and CI-CP, respectively) profiles over $p > b$ binary attributes is NP-complete.*

*Proof.* We only show an argument for the class CI-UP. The reasoning for other two types of profiles is similar. Moreover, we only show that the evaluation problem (under the restriction to profiles consisting of CI-UP trees) is NP-complete.

Indeed, it directly implies that the corresponding variant of the winner problem is NP-hard.

As in other arguments before, the membership in the class NP is evident. Thus, we focus on the hardness part of the argument. To show NP-hardness, we construct a reduction from the evaluation problem under standard Borda when profiles consist of CI-UP trees ($0$-$Borda^{ev}_{CI\text{-}UP}$, for short). That problem is known to be NP-complete [9].

Given an instance $\langle \mathcal{A}, P, l \rangle$ of $0$-$Borda^{ev}_{CI\text{-}UP}$, where $\mathcal{A}$ is a set of $p$ attributes $X_1, \ldots, X_p$, $P = \langle T_1, \ldots, T_m \rangle$ is a profile of $m$ CI-UP trees over $\mathcal{A}$, and $l$ is a positive integer, we construct an instance $\langle \mathcal{A}', P', l' \rangle$ of our problem as follows.

First, we define $\mathcal{A}' = \{Y_1, \ldots, Y_c, X_1, \ldots, X_p\}$, where $Y_1, \ldots, Y_c$ are new attributes. Second, we construct a UI-UP tree $T$ built of $c$ nodes labeled $Y_1, \ldots Y_c$ (from top to bottom), with the node labeled with $Y_i$ having a local preference $1 > 0$. Then, for each $T_i \in P$, $1 \leq i \leq m$, we form a CI-UP tree $T'_i$ by connecting the bottom node of $T$ (the one labeled wit $Y_c$) by a "straight-down" edge to the root of $T_i$. We define $P' = \{T'_1, \ldots, T'_m\}$. Finally, we set $l' = l$.

Similar to the proof of Theorem 4, it is straightforward to verify that under the profile $P$ there is an alternative with the standard Borda score of at least $l$ if and only if under the profile $P'$ there is an alternative with the $b$-Borda score of at least $l'$.                                                                                □

In Table 1, we summarize the results we obtained for $k$-Approval with $k = 2^{p-1} \pm f(p)$ for every polynomial $0 < f(p) < 2^{p-1}$, and for $b$-Borda for every fixed integer $b > 0$. (Whenever the evaluation problem is NP-complete, the corresponding winner problem is NP-hard.)

**Table 1.** Complexity results

|    | UP | CP |
|----|-----|-----|
| UI | P (Thms 1&2) | P (Thms 1&2) |
| CI | P (Thms 1&2) | P (Thms 1&2) |

(a) $(2^{p-1} \pm f(p))$-Approval for $0 < f(p) < 2^{p-1}$

|    | UP | CP |
|----|-----|-----|
| UI | NPC (Thms 3&4) | NPC (Thm 5) |
| CI | NPC (Thm 5) | NPC (Thm 5) |

(b) $b$-Borda for $b > 0$

## 5   Conclusions and Future Work

Aggregating votes specified as LP-trees leads to interesting theoretical and practical problems. In particular, the complexity of the winner and evaluation problems for positional scoring rules is far from being completely understood. First results on the topic were provided by Lang et al. [9,10]. Our results on some of the open problems include new complexity results for special cases of $k$-Approval and for a generalized Borda rule, called $b$-Borda. Specifically, we find that, for all of the four classes of LP-trees (UI-UP, UI-CP, CI-UP, and CI-CP), the winner and evaluation problems are in $P$ under the $k$-Approval rules for $k$ being any integer within a difference of $f(p)$ from $2^{p-1}$, where $f(p)$ is a polynomial of the number of attributes $p$, and that the two problems are NP-hard under $b$-Borda for every $b$ such that $0 < b < p$. However, a full understanding of what makes a positional scoring rule hard remains an open problem.

In the future work, we propose to conduct an empirical study of computational tools (e.g., Maximum Satisfiability and Answer Set Programming solvers) to aggregate votes given in the form of LP-trees.

**Acknowledgments.** The work of the second author was supported by the NSF grant IIS-1618783.

# References

1. Arrow, K.J., Sen, A., Suzumura, K.: Handbook of Social Choice and Welfare, vol. 1. Elsevier, Amsterdam (2002)
2. Booth, R., Chevaleyre, Y., Lang, J., Mengin, J., Sombattheera, C.: Learning conditionally lexicographic preference relations. In: ECAI, pp. 269–274 (2010)
3. Boutilier, C., Brafman, R., Domshlak, C., Hoos, H., Poole, D.: CP-nets: a tool for representing and reasoning with conditional ceteris paribus preference statements. J. Artif. Intell. Res. **21**, 135–191 (2004)
4. Brams, S.J., Fishburn, P.C.: Voting procedures. In: Arrow, K.J., Sen, A.K., Suzumura, K. (eds.) Handbook of Social Choice and Welfare, vol. 1, pp. 173–236. Elsevier, Amsterdam (2002)
5. Saint-Cyr, F.D.D., Lang, J., Schiex, T.: Penalty logic and its link with dempster-shafer theory. In: Proceedings of the Tenth International Conference on Uncertainty in Artificial Intelligence (1994)
6. Dubois, D., Lang, J., Prade, H.: A brief overview of possibilistic logic. In: ECSQARU, pp. 53–57 (1991)
7. Fargier, H., Conitzer, V., Lang, J., Mengin, J., Schmidt, N.: Issue-by-issue voting: an experimental evaluation. In: MPREF (2012)
8. Souhila, K.: Working with Preferences: Less Is More. Cognitive Technologies, Springer, Berlin (2011). https://doi.org/10.1007/978-3-642-17280-9
9. Lang, J., Mengin, J., Xia, L.: Aggregating conditionally lexicographic preferences on multi-issue domains. In: CP, pp. 973–987 (2012)
10. Lang, J., Mengin, J., Xia, L.: Voting on multi-issue domains with conditionally lexicographic preferences. Artif. Intell. **265**, 18–44 (2018)
11. Liu, X., Truszczynski, M.: Aggregating conditionally lexicographic preferences using answer set programming solvers. In: Perny, P., Pirlot, M., Tsoukiàs, A. (eds.) ADT 2013. LNCS (LNAI), vol. 8176, pp. 244–258. Springer, Heidelberg (2013). https://doi.org/10.1007/978-3-642-41575-3_19
12. Liu, X., Truszczynski, M.: Learning partial lexicographic preference trees over combinatorial domains. In: Proceedings of the 29th AAAI Conference on Artificial Intelligence (AAAI), pp. 1539–1545. AAAI Press (2015)
13. Liu, X., Truszczynski, M.: Preference learning and optimization for partial lexicographic preference forests over combinatorial domains. In: Ferrarotti, F., Woltran, S. (eds.) FoIKS 2018. LNCS, vol. 10833, pp. 284–302. Springer, Cham (2018). https://doi.org/10.1007/978-3-319-90050-6_16
14. Schmitt, M., Martignon, L.: On the complexity of learning lexicographic strategies. J. Mach. Learn. Res. **7**, 55–83 (2006)

# On the Parameterized Complexity of Party Nominations

Neeldhara Misra[(⊠)]

Indian Institute of Technology, Gandhinagar, India
neeldhara.m@iitgn.ac.in

**Abstract.** Consider a fixed voting rule r. In the POSSIBLE PRESIDENT problem, we are given an election where the candidates are partitioned into parties, and the problem is to determine if, given a party P, it is possible for every party to nominate a candidate such that the nominee from P is a winner of the election that is obtained by restricting the votes to the nominated candidates. In previous work on this problem, proposed by [10], it was established that POSSIBLE PRESIDENT is NP-hard even when the voting rule is Plurality and the election is restricted to single-peaked votes. In this contribution, we initiate a study of the parameterized complexity of the problem. Our main result is that for a natural choice of parameter (namely the number of parties), the problem is W[2]-hard in general but is FPT on the 1D-Euclidean domain. On the other hand, if we parameterize by the size of the largest party, we encounter para-NP-hardness even on profiles that are both single-peaked and single-crossing. This strengthens previously established hardness results. We also show a polynomial time result for the related NECESSARY PRESIDENT problem on single-crossing elections.

## 1 Introduction

We consider an election scenario where the candidates are split into parties, and every party is asked to nominate a candidate. The votes, while originally preference orders over all alternatives, are now restricted to the nominees and the voting rule is employed on this smaller profile to determine the winner. The decision that the parties now have to make is: which nominee gives them their best shot in the election?

Note that unless all votes rank all candidates from a party consecutively, choices of nominees can lead to significant differences in outcome. For example, assume that the voting rule at play is Plurality, and consider an election where $a \succ x \succ b$ for 5 votes and $b \succ x \succ a$ for 4 votes and $x \succ b \succ a$ for one vote. Suppose the parties were given by $P := \{a, b\}$ and $Q := \{x\}$. Although the election projected on P shows both candidates as being equally competent, it is clear that P should nominate $a$ to (co-)win the election. Beyond toy illustrations, the history of political elections is rich in speculation—if not evidence—concerning the influence of choices of nominations on the final outcomes.

© Springer Nature Switzerland AG 2019
S. Pekeč and K. B. Venable (Eds.): ADT 2019, LNAI 11834, pp. 112–125, 2019.
https://doi.org/10.1007/978-3-030-31489-7_8

There are two natural interpretations for how the parties address this question, depending on how much information they have about the other nominations and their degree of optimism about the information that they do not have. Let P denote an arbitrary but fixed party in this setting. On one extreme, note that if members of P are aware of the nominations made by all the other parties, then the choice for P is straightforward. Indeed, let's say that a candidate c from the party P is *strong* in a vote v if c is ranked higher in v than the highest ranked nominee from the other parties. The potential plurality score of a candidate c is the number votes for which the candidate is strong. The party should nominate any of the candidates that enjoy the highest potential plurality score—this would clearly be their best bet.

On the other hand, assuming access to this type of information is not a realistic starting point, so we now consider the other end of the spectrum, where we have no information about nominees from the other parties. Here, two natural questions arise: the first is to ask if P has any chance at all of winning the election, and the other is to ask if P is guaranteed to win the election. Specifically, the former question asks if there is *any choice of nominees of the other parties* for which the party P has a winning candidate to nominate. In the latter, we ask if P has a winning candidate *no matter which candidates are nominated by the other parties*. The questions, dubbed POSSIBLE PRESIDENT and NECESSARY PRESIDENT respectively, were introduced and studied by [10]. We now turn to a summary of their results. For a broader perspective on questions associated with nominations, uncertainty and strategic behavior, we refer the reader to the discussion in [10] as well as the Handbook of Computational Social Choice [3].

## 1.1 Related Work and Our Results

We briefly outline here the results obtained by [10] to establish the context for this contribution, which can be thought of as a follow-up addressing complementary questions and expanding the scope of some prior results. We refer the reader to the next section for the definitions of the terms used here. In the context of the Pluarlity voting rule, it was established that POSSIBLE PRESIDENT is NP-complete and that NECESSARY PRESIDENT is coNP-complete, even when the size of the largest party is two. These hardness results motivated the study of the problems on restricted domains. It turned out that NECESSARY PRESIDENT was polynomially solvable on single-peaked profiles, while in stark contrast, POSSIBLE PRESIDENT remained NP-complete on 1D-Euclidean profiles, which are a subclass of profiles that are *both* single-peaked and single-crossing.

Finally, it was also established that POSSIBLE PRESIDENT admits a polynomial time algorithm if the elections are restricted to single-peaked profiles where the candidates of any party appear consecutively on the societal axis. Note that this assumption does not force the candidates belonging to the same party to appear consecutively on the *votes*[1], which makes this algorithm an interesting intermediate result.

---

[1] In this case, note that the outcome of the election does not depend on who is nominated.

Given the hardness of POSSIBLE PRESIDENT on restricted domains even for the plurality voting rule, one is compelled to look further into the source of this hardness. There are two counter-balancing elements at play here: the first is the number of parties involved and the other is the sizes of the individual parties. From [10], we already know that the problem remains hard even when all parties have sizes two or less, which is a sharp hardness threshold (note that dealing with parties of size one amounts to asking for winner determination). What about the other extreme, when the number of parties is a constant? Indeed, one might even make the case that this would be a reasonable assumption for many application scenarios, including that of several types of political elections. Note that here we could simply try all possible choice of nominees from the parties and maintain a reasonable running time, and this would work for any voting rule that admits efficient winner determination. We remark here that for the rest of our discussion, we fix Plurality as our choice of voting rule, unless mentioned otherwise.

To dig deeper into the question of whether the problem is easier when the number of parties is small, we invoke the framework of parameterized complexity (we refer readers to [5] for a comprehensive introduction to this approach). In particular, we ask if POSSIBLE PRESIDENT is FPT when parameterized by the number of parties. While we will elaborate shortly on our resolution of this question, we summarize for now by saying that the answer is in the negative for general instances, although we are able to obtain a FPT algorithm for the case of 1D-Euclidean profiles.

Let us return to the issue of small party sizes for a moment. Given that the situation of "many small parties" is hard to handle in general, then it is natural to ask if there is some algorithmic advantage when we assume this situation in the context of restricted domains. We demonstrate here that POSSIBLE PRESIDENT remains NP-complete on profiles that are both single-peaked and single-crossing, even when each party has at most two candidates. Therefore, we continue to witness the sharp threshold on restricted domains as well.

Finally, we also briefly consider the NECESSARY PRESIDENT problem. Recall that this problem is coNP-complete in general but in P for single-peaked profiles (with no requirements for how the candidates appear on the societal axis). As a complementary consideration, we show that the problem is also in P for single-crossing profiles with no additional requirements.

## 1.2   Methodology

We now describe our results more explicitly and briefly overview the tools used to obtain them. For the POSSIBLE PRESIDENT problem in the context of Plurality, we have the following results.

▷ When parameterized by the number of parties (say t), the POSSIBLE PRESIDENT problem is in XP and also W[1]-hard (see Theorem 1).
▷ When restricted to profiles that are 1D-Euclidean, the problem is FPT parameterized by t (see Theorem 2).

▷ When parameterized by the size of the largest party (say $\ell$), the POSSIBLE PRESIDENT problem is para-NP-hard, even when constrained to profiles that are both single-peaked and single-crossing (see Theorem 3).

The first result is obtained by a fairly natural reduction from a variant of dominating set, a standard W[2]-hard problem. The second result is inspired by the dynamic programming used in [10] for the case of profiles that are single-peaked where the candidates that are from the same party appear consecutively on the societal axis. The accuracy of DP approach employed here is driven by both the single-peaked and single-crossing properties of the profiles. The "single-crossing" aspect of 1D-Euclidean profiles allows us say that the votes where the nominee of a party is favored over all others form a consecutive chunk, which gives us a mechanism for anchoring the DP. Exploiting our allowance for a FPT running time, we guess the sequence in which the parties win over the votes (leading to t! possibilities). Finally, the "single-peaked" aspect of the profile, because of which the plurality score of any candidate is influenced only by its immediate neighbors, allows us to formulate an appropriate recurrence. While the algorithmic framework used here is standard, the details do require some attention.

The third result is a reduction from SAT, not unlike the one used by [10] to show NP-hardness on general instances. However, we reduce from a particular variant of SAT that allows us to control the structure of the reduced instance. This variant, called LSAT, was introduced in [1] and has been used recently in other reductions concerning structured profiles [13].

We also consider the NECESSARY PRESIDENT problem briefly, in the context of single-crossing profiles, and obtain a polynomial-time algorithm in this setting. Here we use an overall framework that is standard for such problems, while exploiting the single-crossing property—in a manner similar to what was said of the 1D-Euclidean profiles previously—to obtain an efficient algorithm.

▷ The NECESSARY PRESIDENT problem, when restricted to single-crossing profiles, is in P (see Theorem 4). The analogous result—namely that NECESSARY PRESIDENT is in P when restricted to single-peaked profiles—was shown in [10].

## 2  Preliminaries

In this section, we summarize the definitions and notations that we will need subsequently. We refer the reader to [3] for a comprehensive introduction to the study of elections in a computational setting.

### 2.1  Notations and Definitions

For a positive integer $\ell$, we denote the set $\{1, \ldots, \ell\}$ by $[\ell]$. We first define some general notions relating to voting rules. An election comprises of candidates and voters. Let $V = \{v_1, \ldots, v_n\}$ be a set of $n$ *voters* and $C = \{c_1, \ldots, c_m\}$ be a set of m *candidates*. If not mentioned otherwise, we denote the set of candidates, the set of voters, the number of candidates, and the number of voters by $C$, $V$, $m$, and $n$ respectively.

Every voter $v_i$ has a *preference* $\succ_i$ which is a complete order over the set $C$ of candidates. We say voter $v_i$ prefers a candidate $x \in C$ over another candidate $y \in C$ if $x \succ_i y$. We denote the set of all preferences over $C$ by $\mathscr{L}(C)$. The $n$-tuple $(\succ_i)_{i \in [n]} \in \mathscr{L}(C)^n$ of the preferences of all the voters is called a *profile*. Note that a profile, in general, is a multiset of linear orders. For a subset $M \subseteq [n]$, we call $(\succ_i)_{i \in M}$ a sub-profile of $(\succ_i)_{i \in [n]}$. For a subset of candidates $D \subseteq C$, we use $\mathscr{P}|_D$ to denote the projection of the profile $\mathscr{P}$ on the candidates in $D$ alone. We abuse notation and use $V$ to denote both the set of voters and the profile associated with them, whenever the distinction is clear from the context.

A *domain* is a set of profiles. Below, we describe what it means for a profile to belong to certain well-studied domains. As a matter of terminology, and usually in the context of computational problems, we often say that an election has a certain property (e.g, *the election is single-peaked*) to refer to the fact that the voter profile of that election instance belongs to the corresponding domain. We refer the reader to the recent survey of [8] for a detailed overview of structured preferences.

*Single-Peaked.* A preference profile is said be single-peaked if there exists an ordering $\sigma$ over the candidates $C$ such that the preference of every voter $v$ has the following structure: $v$ has a favorite candidate $c$ (sometimes called the "peak" for $v$), and the further away a candidate $d \neq c$ is from $c$ in $\sigma$, the less it is preferred by the voter $v$. The notion of single-peaked preferences was introduced by [2] and a formal definition is as follows. It is well-known that it is possible to check if a given election is single-peaked in polynomial time (see, for instance, [9]).

**Definition 1 (Single-peaked).** *A preference $\succ \in \mathscr{L}(C)$ over a set of candidates $C$ is called* single-peaked *with respect to an order $\succ' \in \mathscr{L}(C)$ if, for every pair of candidates $x, y \in C$, we have $x \succ y$ whenever we have either $c \succ' x \succ' y$ or $y \succ' x \succ' c$, where $c \in C$ is the candidate at the first position of $\succ$. A profile $\mathscr{P} = (\succ_i)_{i \in [n]}$ is called single-peaked with respect to an order $\succ' \in \mathscr{L}(C)$ if $\succ_i$ is single-peaked with respect to $\succ'$ for every $i \in [n]$.*

*Single-Crossing.* A preference profile is said to belong to the single-crossing domain if it admits a permutation of the voters such that for any pair of candidates $a$ and $b$, there is an index $j[(a,b)]$ such that either all voters $v_j$ with $j < j[(a,b)]$ prefer $a$ over $b$ and all voters $v_j$ with $j > j[(a,b)]$ prefer $b$ over $a$, or vice versa. The notion was introduced in [12,14] The formal definition is as follows.

**Definition 2 (Single-crossing).** *A profile $\mathscr{P} = (\succ_i)_{i \in [n]}$ of $n$ preferences over a set $C$ of candidates is called a* single-crossing *profile if there exists a permutation $\sigma$ of $[n]$ such that, for every pair of distinct candidates $x, y \in C$, whenever we have $x \succ_{\sigma(i)} y$ and $x \succ_{\sigma(j)} y$ for two integers $i$ and $j$ with $1 \leqslant \sigma(i) < \sigma(j) \leqslant n$, we have $x \succ_{\sigma(k)} y$ for every $\sigma(i) \leqslant k \leqslant \sigma(j)$.*

*1D-Euclidean.* A preference profile is said to belong to the 1D-Eucidean domain if the voters and candidates can be arranged on a line such that the preference of any voter $v$ is determined by the distances of the candidates from the $v$: in particular, $v$ prefers alternative c to d if the distance of $v$ to c is smaller than the distance to d. Profiles that are 1D-Euclidean are single-peaked and single-crossing, but the converse is not true [4,7]. 1D-Euclidean profiles can also be recognized in polynomial time [6,11].

**Definition 3 (1D-Euclidean).** *A profile $\mathscr{P} = (\succ_i)_{i \in [n]}$ of $n$ preferences over a set C of candidates is called a* 1D-Euclidean single-peaked profile *if there exists a function $f : C \cup [n] \to \mathbb{R}$ such that for each voter $t \in [n]$ and each pair of candidates $c_i, c_j \in C$, it holds that $c_i \succ_t c_j$ if and only if $|f(v_t) - f(c_i)| < |f(v_t) - f(c_j)|$.*

## 2.2  Problem Definitions

We now formally define the notions of a possible and necessary president, and the natural computational problems associated with them. These definitions were introduced in [10]; we reproduce them below for completeness.

**Definition 4.** *Let r be a fixed voting rule, and let $E = (C, V)$ be an election where we are also given a partition of the set of candidates C into parties $\mathscr{P} = \{P_1, \ldots, P_t\}$. A subset of candidates $N \subseteq C$ is called a set of nominees if $|N \cap P_i| = 1$ for all $i \in [t]$.*

▷ *A party $P_w$ is said to have an* possible president *if there is a set of nominees $N \subseteq C$ such that $r(N, V|_N)$ contains a member of $P_w$.*
▷ *A party $P_w$ is said to have an* necessary president *if there is a candidate $c \in P_w$ such that for any set of nominees N which contains c, $r(N, V|_N)$ contains c.*

If a party $P_w$ has only one candidate, say p, then we abuse terminology and say that p is a possible (respectively, necessary) president to refer to the fact that $P_w$ has a possible (respectively, necessary) president. We now turn to the formulation of r-POSSIBLE PRESIDENT, which is the computational problem associated with the notion of a possible president.

---

r-POSSIBLE PRESIDENT
**Input:** An election $E = (C, V)$ along with a split of C into t parties $P = \{P_1, \ldots, P_t\}$, and an integer $w \in [t]$.
**Question:** Does $P_w$ admit a possible president?

---

The r-NECESSARY PRESIDENT problem is defined analogously. We use $\ell$ to denote the size of the largest party in an instance of either problem, while t denotes the number of parties in the input. We focus entirely on the Plurality voting rule where the winners are the candidates ranked on the first place by the largest number of voters. The score of a candidate c is the number of voters that rank it first. We assume the nonunique-winner model, so if there are multiple candidates with the highest score, then we declare all of them as winners.

# 3  Possible President

In this section, we focus on Plurality-POSSIBLE PRESIDENT. First, we establish some results about the parameterized complexity of the problem on general domains when parameterized by the number of parties.

## 3.1  Parameterized Complexity Results

In this subsection we establish our results for Plurality-POSSIBLE PRESIDENT when parameterized by the number of parties. For the W[1]-hardness result, we will reduce from MULTICOLORED RED-BLUE DOMINATING SET (abbreviated M-RBDS), which is well-known to be a complete problem for W[2] (see [5]) and is defined as follows. The input is a bipartite graph $G = (V = (R \cup B), E)$ and a partition of the vertex set R into k parts given by $R = \uplus_{i=1}^{k} R_i$. The question is if G admits a set S of size k whose intersection with $R_i$ is exactly one for all $i \in [k]$ and is such that every vertex in B has a neighbor in S.

**Theorem 1.** *The Plurality-*POSSIBLE PRESIDENT *problem admits an algorithm with running time* $O^*(m^t)$ *and is W[1]-hard when parameterized by* t.

*Proof.* Towards the algorithm, guess the set of nominees. If we use $m_i$ to denote the number of candidates in the party $P_i$, then the running time of this brute-force approach is given by $O(m^t)$ using the standard relationship between arithmetic geometric means. To establish W[1]-hardness, we reduce from MULTICOLORED RED-BLUE DOMINATING SET. Consider an instance of M-RBDS given by:

$$(G = ((R \cup B), E); R = R_1 \uplus R_2 \uplus \cdots \uplus R_t; t).$$

Let $R := \{r_1, \ldots, r_m\}$ and $B := \{b_1, \ldots, b_n\}$. We now describe the reduced instance of Plurality-POSSIBLE PRESIDENT. For each vertex $r_i \in R$, introduce a candidate $c_i$. Let the parties be given by the partition of the M-RBDS instance—in other words, for each $j \in [t]$, all candidates corresponding to vertices in $R_j$ form the party $P_j$. Also introduce two new candidates $p$ and $q$, both of whom form their own (singleton) parties. We use C to denote the set of candidates corresponding to vertices in R. Let $\sigma$ be an arbitrary but fixed ordering over C. For a subset $D \subseteq C$, the sequence given by elements of D ordered according to $\sigma$ is denoted by $\overrightarrow{D}$.

We now describe the votes. For each $k \in [n]$, let $S_k$ denote the subset of candidates corresponding to vertices in $N(b_k)$, and introduce the vote $v_k$ given by:

$$v_k : \overrightarrow{S_k} \succ q \succ \overrightarrow{C \setminus S_k} \succ p.$$

Further, define the votes $g_p$ and $g_q$ as:

$$g_p : p \succ q \succ \overrightarrow{C} \text{ and } g_q : q \succ p \succ \overrightarrow{C}.$$

The reduced instance comprises of the votes $v_j$ for $j \in [n]$, and $n$ copies each of the votes $g_p$ and $g_q$. We ask if $p$ is a possible president. The proof of equivalence is fairly straightforward from the construction and is similar in spirit to the proof of NP-hardness shown in [10]. We defer the details here to a full version of this paper. □

We now turn to the case of 1D-Euclidean preferences. Recall that in this scenario, the voters and candidates can be arranged on a line such that the preference of any voter $v$ is given by set of candidates sorted in increasing order of distances from $v$. Recall that Plurality-POSSIBLE PRESIDENT is known to be NP-hard even in this restricted setting. The next result claims that, on the other hand, the problem is FPT when parameterized by the number of candidates. Before the formal description of our algorithm, we sketch the main ideas involved and also introduce some useful terminology.

Recall that an instance of Plurality-POSSIBLE PRESIDENT is given by $(E = (C, V); C = (P_1 \uplus \cdots \uplus P_t); w)$. Let f denote the 1D-Euclidean embedding of $C \cup V$ on $\mathbb{R}$. Let the images of $C \cup V$ under f be given by:

$$\jmath_f := x_1 \leqslant x_2 \leqslant \ldots \leqslant x_{n+m}.$$

For the sake of discussion, let $N \subseteq C$ be a set of nominees and let $\jmath_N$ denote the ordering of $N$ that is consistent with $\jmath_f$. Now, let $\jmath_N^P$ denote the sequence of parties corresponding to candidates in $\jmath_N$: in other words, if the $i^{th}$ nominee belongs to $P_j$, then the $i^{th}$ element of $\jmath_N^P$ is $P_j$. We call $\jmath_N^P$ the *party order of N*. Our algorithm will essentially guess this order (note that there are $t!$ possibilities). The party order helps anchor a dynamic program which is similar (in spirit) to the approach proposed in [10] for single-peaked preferences under the assumption that the candidates of any party appear consecutively on the societal axis.

Now, we turn to an observation about the behavior of a solution in the 1D-Euclidean setting. For a nominee $c \in N$, let $V_c$ be the set of voters for whom $c$ is preferred over any other nominee in $N$. Note that the plurality score of $c$ in the election $(C, V|_N)$ will be $|V_c|$. For a party $P_j$, let $v_{j_L}$ and $v_{j_R}$ denote the leftmost and rightmost voters from $V_c$ with respect to f, where $c$ is the nominee from $P_j$. We claim that if $v_j$ is a voter that is sandwiched between $v_{j_L}$ and $v_{j_R}$ in the embedding, then $v$ must also prefer $c$ over any other nominee in $N$. This follows from the single-crossing nature of the voter order derived from the embedding. We formalize this claim below.

*Claim.* If $v_j \in V$ is such that $f(v_{j_L}) \leqslant f(v_j) \leqslant f(v_{j_R})$, then $v_j \in V_c$.

*Proof.* Indeed, suppose not. Let $d \in N$ be such that $v$ prefers $d$ over $c$. Combined with the assumptions in the claim, we have that:

$$c \succ_{j_L} d, d \succ_j c \text{ and } c \succ_{j_R} d,$$

contradicting the single-crossing nature of the profile—recall that the voters are single-crossing in the order of their appearance in the 1D-Euclidean embedding. □

The claim above motivates the following definition. We say that the *influence of a party* $P_j$ is given by the interval $(L_j, R_j)$ where:

$$L_j = f(v_{j_L}) \text{ and } R_j = f(v_{j_R}).$$

Also, if $c$ is the nominee of a party $P$, then we refer to $|V_c|$ as the strength of the party $P$. We are now ready to provide a high-level description of our approach. We begin by guessing the nominee of the party $P_w$, the influence of $P_w$, and the party order of the solution. For $k \in [n + m]$, we introduce the following notation:

▷ Let $\mathcal{F}_k = (C_k, V_k)$ denote the restriction of the election $(C, V)$ to the first $k$ entities in $\mathfrak{s}_f$, that is:

$$C_k = C \cap \{f^{-1}(x_i) \mid 1 \leqslant i \leqslant k\};$$

$$V_k = V \cap \{f^{-1}(x_i) \mid 1 \leqslant i \leqslant k\}.$$

▷ Analogously, let $\mathcal{F}_k = (C_k', V_k')$ denote the restriction of the election $(C, V)$ to the last $(m - k + 1)$ entities in $\mathfrak{s}_f$.

Suppose $c^\star \in P_w$ is the guessed nominee from $P_w$, and let $(L_w, R_w)$ denote the assumed influence of $P_w$. We use $S$ to denote the strength of $P_w$—note that this is determined once the influence is fixed. Let $\mathfrak{s}^P$ denote the guessed party order—by renaming, assume that this is order is given by $P_1, P_2, \ldots, P_t$. We will also assume, for ease of discussion, that $w$ is neither 1 nor $t$, noting that it is straightforward to handle these "edge cases" separately. We now ask if it is possible to choose nominees $N$ such that:

▷ $c \in N$,
▷ For any $i, j \in [t]$ such that $i < j$, $f(c_i) < f(c_j)$, where $c_i$ and $c_j$ denote the nominees of parties $P_i$ and $P_j$, respectively, and
▷ The strength of any party is at most $S$.

Let $P_L$ (respectively, $P_R$) denote the parties that occur to the left (respectively, right) of $P_w$. Note that the second condition above implies that for a party $P \in P_L$, the nominee of $P$ belongs to $C_{L_w - 1}$ and similarly, for a party $P \in P_R$, the nominee of $P$ belongs to $C_{R_w + 1}$. We answer the question above separately for the elections $\mathcal{F}_{L_w - 1}$ and $\mathcal{F}_{R_w + 1}$. Our discussion will focus on resolving the instance $\mathcal{F}_{L_w - 1}$. The details for $\mathcal{F}_{R_w + 1}$ are analogous. We use dynamic programming and begin by specifying the entries and semantics of the table. For each $1 \leqslant i, \ell_1, \ell_2, \leqslant L_w - 1$, and $1 \leqslant j \leqslant w - 1$, define:

$$\mathbb{T}[k, i, \ell_1, \ell_2] = \begin{cases} 1 \text{ if } (\star) \text{ holds,} \\ 0 \text{ otherwise,} \end{cases}$$

where $(\star)$ states that there exists nominees $d_1, \ldots, d_i$ from parties $P_1, \ldots, P_i$, each with strength at most $S$, such that $f(d_1) < f(d_2) < \cdots < f(d_i)$, and further:

$$x_{\ell_1} \leqslant f(d_i) \leqslant x_{\ell_2}.$$

We expect that the reader will find this definition to be mostly along expected lines, possibly except for the last condition. We briefly describe the intuition for the requirement imposed on $f(d_i)$. To this end, note that the standard way to compute $\mathbb{T}[k, i, \ell_1, \ell_2]$ would be by a recursive formulation which exhaustively examines all reasonable possibilities for the nominee of the party $P_i$ and its influence, and then attempt to "patch" the choice made with the solution of an appropriate subproblem. However when we invoke a recursively smaller subproblem, we need to "remember" that the influence of the party $P_i$ has ended, and in particular, we need to ensure that we choose a candidate close enough to voters on the extreme right. In other words, for the subproblem that we appeal to when we recurse, we will ensure that voters do not find the chosen candidate from $P_i$ (now no longer represented in the instance) closer to them compared to the candidate we are going to choose from $P_{i-1}$. By insisting that the candidate chosen from $P_{i-1}$ is to the right of $x_{\ell_1}$, we can enforce this by a careful choice of $x_{\ell_1}$. The requirement for staying to the left of $x_{\ell_2}$ is analogous: here, the intent is to guard the influence of $P_i$ (or equivalently, prevent an unintended extension of the influence of $P_{i-1}$). Specifically, this condition will help us with ensuring that the candidate chosen from $P_{i-1}$ does not influence some of the deleted voters.

Now, recall that $L_w$ is the assumed extent of the leftward influence of the party $P_w$. Let $d_1$ denote the distance between $c^\star$ and the rightmost voter on the 1D-Euclidean ordering that is to the left of $L_w$. Further, among all candidates who are embedded to the right of $f(c^\star) - 2d_1$, let $c'$ denote the leftmost and let $f(c') = x_{\ell_1}$. Similarly, let $d_2$ denote the distance between $c^\star$ and the leftmost voter to the right of $L_w$. Among all candidates who are embedded to the left of $f(v) - d_2$, let $c''$ denote the rightmost and let $f(c'') := x_{\ell_2}$. Then observe that we have a solution to the left half of our problem if and only if $\mathbb{T}[L_w - 1, w - 1, \ell_1, \ell_2] = 1$.

We are now ready to describe the recursive formulation for $\mathbb{T}[k, i, \ell_1, \ell_2]$. When $i = 1$ we compute the values of $\mathbb{T}$ directly any combination of values for $k$ and $\ell_1, \ell_2$. These entries constitute the base case of the recursion. Observe that these values are easy to compute: the strength of any nominee from $P_1$ is equal to the number of voters in the instance $\mathcal{J}_k$, and it is therefore straightforward to check if the strength of $P_1$ exceeds $S$ or not. Since any candidate from $P_1$ has the same strength, the conditions associated with $\ell_1, \ell_2$ is merely checked by inspection: so $\mathbb{T}[k, 1, \ell_1, \ell_2]$ resolves to 1 if there exists some $c \in P_1$ such that $x_{\ell_1} \leqslant f(c) \leqslant x_{\ell_2}$ and $|V_k| \leqslant S$.

Now consider $\mathbb{T}[k, i, \ell_1, \ell_2]$ for $i \geqslant 2$. We use $Q_i$ to denote the set of all legitimate choices for a nominee from the party $P_i$. Note that a candidate $q \in P_i$ belongs to $Q_i$ if and only if $x_{\ell_1} \leqslant f(q) \leqslant x_{\ell_2}$ and $q \in C_k$. If $Q_i = \emptyset$, then $\mathbb{T}[k, i, \ell_1, \ell_2] = 0$. Otherwise, for $q \in Q_i$, let $x_{j_q} := f(q)$ and let $j'_q$ be the smallest index for which $f^{-1}(x_{j'}) \in V$ and the number of voters mapped to the right of $x_{j'_q}$ (inclusive) is at most $S$. Intuitively, $j'_q$ denotes the furthest permissible extent of the influence of $P_i$. We claim that the following recurrence holds:

$$\mathbb{T}[k, i, \ell_1, \ell_2] = \bigvee_{q \in Q_i} \mathbb{T}[k - j'_q, i - 1, \ell_1^q, \ell_2^q],$$

where the values of $\ell_1^q, \ell_2^q$ are chosen appropriately based on $f(q)$ and $j'_q$. Note that the computation of $\mathbb{T}[k, i, \ell_1, \ell_2]$ depends on values of $\mathbb{T}$ for smaller $i$, making the recurrence well-defined when combined with the base case above. We are now ready to summarize this discussion in the following result.

**Theorem 2.** *The Plurality-*POSSIBLE PRESIDENT *is FPT when parameterized by* t *when the input election is 1D-Euclidean.*

*Proof (Sketch).* Our algorithm examines all possible choices of the nominee of the party $P_w$, the influence of $P_w$, and the party order of the solution. Note that this gives rise to at most $O(mn^2t!)$ possible scenarios to handle. For each guess, we proceed by employing the dynamic programming approach outlined above. It is easy to see that the DP tables have polynomially many entries and each entry can be updated in constant time. The correctness follows from the fact that the recurrence proposed above is an exhaustive examination of all possibilities. Owing to space constraints, we defer a formal proof of correctness to the full version of this paper.                                                                 □

### 3.2   The Case of Small Parties

In this section, we show the hardness of computing a possible president on profiles that are 1D-Euclidean (and consequently both single-peaked and single-crossing), even when the size of each party is at most two. This strengthens the NP-hardness result shown in [10] for 1D-Euclidean preferences, although the hard instances constructed there involved large parties. The work of [10] also had a separate reduction for parties of size two by a reduction from SATISFI-ABILITY but with no apparent structure. We perform a reduction in the same spirit but by starting from a structured variant of SATISFIABILITY to achieve the 1D-Euclidean representation.

The variant of SAT we alluded to above is called LINEAR SAT (abbreviated LSAT). In an LSAT instance, each clause has at most three literals, and further the literals of the formula can be sorted such that every clause corresponds to at most three consecutive literals in the sorted list, and each clause shares at most one of its literals with another clause, in which case this literal is extreme in both clauses. The hardness of LSAT was shown in [1]. In fact, by studying the reduced instance, one may assume that a "hard" instance of LSAT has the following structure: the first 2q clauses have two literals each and are of the following form:

$$A_i = \{s_i, \ell_i\}, B_i = \{\ell_i, t_i\}; 1 \leqslant i \leqslant q,$$

where $s_i$, $\ell_i$, and $t_i$ denote literals, while the remaining p clauses (denoted by $C_1, \ldots, C_p$) have three literals each and are mutually disjoint from each other as well as the first 2q clauses. For ease of description, we will assume that the LSAT formula that we reduce from has this particular structure. We are now ready to describe our reduction.

**Theorem 3.** *The Plurality-*POSSIBLE PRESIDENT *problem remains NP-complete on 1D-Euclidean profiles, even when every party has at most two members.*

*Proof.* Let $\phi$ be an instance of LSAT as described above. We first describe the construction of the election for to $\phi$, then argue the equivalance, and finally demonstrate that the suggested profile admits a 1D-Euclidean representation.

First, let us introduce the candidate c, whom we refer to as the *contender* candidate. The contender candidate belongs to a party where she is the sole member. There are exactly two voters who place c at the top. The ordering of the remaining candidates can be arbitrary.

For every clause $A_i$, $B_i$, and $C_j$, we introduce candidates $\alpha_i$, $\beta_i$ and $\gamma_j$ (note that $i \in [q]$ and $j \in [p]$). We refer to these candidates as *rivals*, and every rival candidate belongs to a party where she is the sole member. For every rival candidate, we will introduce exactly two voters that has the rival candidate as his top preference. The ordering of the remaining candidates can be arbitrary.

We also introduce a candidate for every literal (we refer to these as *guards*) and a voter for every clause. For every variable x of $\phi$, we let the candidates corresponding to x and $\overline{x}$ form the party $P_x$. The voter corresponding to any clause places the candidates corresponding to literals that belong to C at the top of his preference list in an arbitrary order immediately followed by the rival candidate corresponding to the clause. The ordering of the remaining candidates can be arbitrary. This completes the construction of the instance. The spirit of our argument for equivalance is similar to the one in [10].

In the forward direction, let $\tau$ be a satisfying assignment for $\phi$. Nominate from each party $P_x$ the variable that was set to TRUE by $\tau$. All other parties are singletons so their nominations are forced. Now consider the Plurality scores of the candidates. The score of the contender is exactly three, since all voters other than the ones who placed her on top have rivals preferred over the contenders. The score of any guard is at most two, since every literal appears in at most two clauses, and the plurality score of a nominated guard is equal to the number of clauses that she appears in. It is easily verified that the score of any rival candidate is exactly two.

In the reverse direction, let N be a fixed choice of nominations. Observe that irrespective of the nature of N, the contender has a final Plurality score of exactly two, and every rival candidate has a Plurality score of at least two. We now define $\tau$ as the assignment that mimics the choices of nominations from parties corresponding to variables: if x is nominated from $P_x$, $\tau(x) = 1$ and if $\overline{x}$ is nominated from $P_x$, then $\tau(x) = 0$. We claim that $\tau$ is a satisfying assignment for $\phi$. Suppose not, and in particular, let C be a clause that is not satisfied. Then, by construction, the voter corresponding to C contributes one to the Plurality score of the rival candidate corresponding to C, preventing the contender candidate from being a cowinner.

We now turn to an informal description of the underlying 1D-Euclidean structure of the constructed instance. We place the literals corresponding to the clauses $C_i$'s in close clusters that are far apart from each other, and we also place w at a point far away from the clause clusters. We place the clauses corresponding to the $A_i$'s and $B_i$'s as shown in Fig. 1, and it is easily verified that one can come up with a placement of voters that gives us an instance with all the properties demanded by the construction described above.                      □

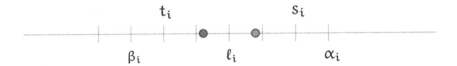

**Fig. 1.** A 1D-Euclidean representation for the clauses $A_i$ and $B_i$. Observe that a voter placed at the blue circle on the left would have $\ell_i \succ t_i \succ \beta_i$ as his top three candidates, while a voter placed at the red circle on the right would have $\ell_i \succ s_i \succ \alpha_i$ as his top three preferences, as desired.

## 4    Necessary President

In this section, we switch our focus to the Plurality-NECESSARY PRESIDENT problem. It is established that the problem is coNP-complete in general and also in P on single-peaked profiles. We address the question of the complexity of Plurality-NECESSARY PRESIDENT on single-crossing profiles, and show that the problem admits a polynomial-time algorithm in this setting as well.

**Theorem 4** ($\star$). *Plurality-*NECESSARY PRESIDENT *is in P if the input election is single-crossing.*

For lack of space, we provide here an informal summary of the main ideas in lieu of a detailed proof. As is typical for these problems, our strategy is to guess a nominee $c$ from the party $P_w$ and a rival candidate $d$, which allows us to focus on the following question: is there a set of nominees that ensures that the final plurality score of $d$ exceeds that of $c$? Towards addressing this question, we show an analog of Claim 3.1 for single-crossing profiles. In particular, it is easy to show the following: if $N$ is a set of nominees and $c \in N$, then the set of voters that rank $c$ higher than any of the other nominees form a consecutive block in the single-crossing order. Given this, we can also guess the start and end points of the blocks corresponding to $c$ and $d$. After suitable sanity checks, what we now need to ensure is the following: for any vote $v$ that does not appear in the blocks corresponding to $c$ or $d$, we must choose at least one valid candidate among those that are ranked higher than $c$ by $v$ (a candidate is valid if it is not ranked higher than $c$ or $d$ in their blocks). This is expressed naturally as a hitting set type requirement, but is easily checked because of the very special nature of the profile and the restrictions imposed by the guesses made so far.

## 5    Concluding Remarks

Our contribution offers a parameterized complexity perspective on the Plurality-POSSIBLE PRESIDENT problem. We state here two pertinent directions. Is Plurality-POSSIBLE PRESIDENT FPT when parameterized by $t$ on single-peaked domains, or single-crossing domains, or domains that are both single-peaked and single-crossing? What is the parameterized complexity of the problem when parameterized by the number of voters?

# References

1. Arkin, E.M., Banik, A., Carmi, P., Citovsky, G., Katz, M.J., Mitchell, J.S.B., Simakov, M.: Choice is hard. In: Elbassioni, K., Makino, K. (eds.) ISAAC 2015. LNCS, vol. 9472, pp. 318–328. Springer, Heidelberg (2015). https://doi.org/10.1007/978-3-662-48971-0_28
2. Black, D.: On the rationale of group decision-making. J. Polit. Econ. **56**, 23–34 (1948)
3. Brandt, F., Conitzer, V., Endriss, U., Lang, J., Procaccia, A.: Handbook of Computational Social Choice (2016)
4. Chen, J., Pruhs, K., Woeginger, G.J.: The one-dimensional euclidean domain: finitely many obstructions are not enough. Soc. Choice Welf. **48**(2), 409–432 (2017)
5. Cygan, M., et al.: Parameterized Algorithms. Springer, Cham (2015). https://doi.org/10.1007/978-3-319-21275-3
6. Elkind, E., Faliszewski, P.: Recognizing 1-euclidean preferences: an alternative approach. In: Lavi, R. (ed.) SAGT 2014. LNCS, vol. 8768, pp. 146–157. Springer, Heidelberg (2014). https://doi.org/10.1007/978-3-662-44803-8_13
7. Elkind, E., Faliszewski, P., Skowron, P.: A characterization of the single-peaked single-crossing domain. In: Proceedings of the Twenty-Eighth AAAI Conference on Artificial Intelligence, pp. 654–660 (2014)
8. Elkind, E., Lackner, M., Peters, D.: Structured preferences. Trends Comput. Soc. Choice, 187–207 (2017)
9. Escoffier, B., Lang, J., Öztürk, M.: Single-peaked consistency and its complexity. In: Proceedings of the 18th European Conference on Artificial Intelligence (ECAI), pp. 366–370 (2008)
10. Faliszewski, P., Gourvès, L., Lang, J., Lesca, J., Monnot, J.: How hard is it for a party to nominate an election winner? In: Kambhampati, S. (ed.) Proceedings of the Twenty-Fifth International Joint Conference on Artificial Intelligence, IJCAI, pp. 257–263. IJCAI/AAAI Press (2016)
11. Knoblauch, V.: Recognizing one-dimensional euclidean preference profiles. J. Math. Econ. **46**(1), 1–5 (2010)
12. Mirrlees, J.A.: An exploration in the theory of optimum income taxation. Rev. Econ. Stud. **38**(2), 175–208 (1971)
13. Misra, N., Sonar, C., Vaidyanathan, P.R.: On the complexity of chamberlin-courant on almost structured profiles. In: Rothe, J. (ed.) ADT 2017. LNCS (LNAI), vol. 10576, pp. 124–138. Springer, Cham (2017). https://doi.org/10.1007/978-3-319-67504-6_9
14. Roberts, K.W.S.: Voting over income tax schedules. J. Public Econ. **8**(3), 329–340 (1977)

# Gradient Methods for Solving Stackelberg Games

Roi Naveiro[✉] and David Ríos Insua

Institute of Mathematical Sciences (ICMAT-CSIC), Madrid, Spain
roi.naveiro@icmat.es

**Abstract.** Stackelberg Games are gaining importance in the last years due to the raise of Adversarial Machine Learning (AML). Within this context, a new paradigm must be faced: in classical game theory, intervening agents were humans whose decisions are generally discrete and low dimensional. In AML, decisions are made by algorithms and are usually continuous and high dimensional, e.g. choosing the weights of a neural network. As closed form solutions for Stackelberg games generally do not exist, it is mandatory to have efficient algorithms to search for numerical solutions. We study two different procedures for solving this type of games using gradient methods. We study time and space scalability of both approaches and discuss in which situation it is more appropriate to use each of them. Finally, we illustrate their use in an adversarial prediction problem.

**Keywords:** Game theory · Adversarial machine learning · Adjoint method · Automatic differentiation

## 1 Introduction

Over the last decade, the introduction of machine learning applications in numerous fields has grown tremendously. In particular, applications in security settings have grown substantially, [20]. In this domain, it is frequently the case that the data distribution at application time is different of the training data distribution, thus violating one of the key assumptions in machine learning. This difference between training and test distributions generally comes from the presence of adaptive adversaries who deliberately manipulate data to avoid being detected.

The field of Adversarial Machine Learning (AML) studies, among other things, how to guarantee the security of machine learning algorithms against adversarial perturbations [2]. A possible approach consists of modelling the interaction between the learning algorithm and the adversary as a game in which

R. Naveiro acknowledges the Spanish Ministry for his grant FPU15-03636. The work of D.R. Insua is supported by the Spanish Ministry program MTM2017-86875-C3-1-R and the AXA-ICMAT Chair on Adversarial Risk Analysis. D.R. Insua also acknowledges the support of the EU's Horizon 2020 project 815003-2 TRUSTONOMY and the RTC-2017-6593-7 project.

S. Pekeč and K. B. Venable (Eds.): ADT 2019, LNAI 11834, pp. 126–140, 2019.
https://doi.org/10.1007/978-3-030-31489-7_9

one agent controls the predictive model parameters while the other manipulates input data. Several different game theoretic models of this problem have been proposed, as reviewed in [28]. In particular, [5] view adversarial learning as a Stackelberg game in which, a *leader* (she), the defender in the security jargon, makes her decision about choosing the parameters in a learning model, and, then, the *follower* or attacker (he), after having observed the leader's decision, chooses an optimal data transformation.

Mathematically, finding Nash equilibria of such Stackelberg games requires solving a bilevel optimization problem, which, in general cannot be undertaken analytically, [26], and numerical approaches are required. However, standard techniques are not able to deal with continuous and high dimensional decision spaces, as those appearing in AML applications.

In this paper, we propose two procedures to solve Stackelberg games in the new paradigm of AML and study their time and space scalability. In particular, one of the proposed solutions scales efficiently in time with the dimension of the decision space, at the cost of more memory requirements. The other scales well in space, but requires more time. The paper is organized as follows: in Sect. 2 we define Stackelberg games. Section 3 presents the proposed solution methods as well as a discussion of the scalability of both approaches. The proposed solutions are illustrated with an AML experiment in Sect. 4. Finally, we conclude and present some lines for future research.

## 2   Stackelberg Games

We consider a class of sequential games between two agents: the first one makes her decision, and then, after having observed the decision, the second one implements his response. These games have received various names in the literature including sequential Defend-Attack [4] or Stackelberg [10, 27] games. As an example, consider adversarial prediction problems, [5]. In them, the first agent chooses the parameters of a certain predictive model; the second agent, after having observed such parameters, chooses an optimal data transformation to fool the first agent as much as possible, so as to obtain some benefit.

As we focus on applications of Stackelberg games to AML, we restrict ourselves to the case in which the Defender ($D$) chooses her defense $\alpha \in \mathbb{R}^n$ and, then, the Attacker ($A$) chooses his attack $\beta \in \mathbb{R}^m$, after having observed $\alpha$. The corresponding bi-agent influence diagram, [11], is shown in Fig. 1. The dashed arc between nodes $D$ and $A$ reflects that the Defender choice is observed by the Attacker. The utility function of the Defender, $u_D(\alpha, \beta)$, depends on both, her decision, and the attacker's decision. Similarly, the Attacker's utility function has the form $u_A(\alpha, \beta)$. In this type of games, it is assumed that the Defender knows $u_A(\alpha, \beta)$. This assumption is known as the common knowledge hypothesis.

Mathematically, finding Nash equilibrium of Stackelberg games requires solving a bilevel optimization problem, [1]. The defender's utility is called *upper level* or *outer* objective function while the attacker's one is referred to as *lower level* or *inner* objective function. Similarly, the upper and lower level optimization

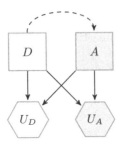

**Fig. 1.** The two-player sequential decision game with certain outcome.

problems, correspond to the defender's and the attacker's problem, respectively. These problems are also referred to as outer and inner problems.

It is generally assumed that the attacker will act rationally in the sense that he will choose an action that maximizes his utility, [8], given the disclosed defender's decision $\alpha$. Assuming that there is a unique global maximum of the attacker's utility for each $\alpha$, and calling it $\beta^*(\alpha)$, a Stackelberg equilibrium is identified using backward induction: the defender has to choose $\alpha^*$ that maximizes her utility subject to the attacker's response $\beta^*(\alpha)$. Mathematically, the problem to be solved by the defender is

$$\begin{aligned} \arg\max_{\alpha} \quad & u_D[\alpha, \beta^*(\alpha)] \\ \text{s.t.} \quad & \beta^*(\alpha) = \arg\max_{\beta} u_A(\alpha, \beta). \end{aligned} \quad (1)$$

The pair $(\alpha^*, \beta^*(\alpha^*))$ is a Nash equilibrium and a sub-game perfect equilibrium [14].

When the attacker problem has more than one global maximum, several types of equilibrium have been proposed. The two more important are the optimistic and the pessimistic solutions, [26]. In an optimistic position, the defender expects the attacker to choose the optimal solution which gives the higher upper level utility. On the other hand, the pessimistic approach suggests that the defender should optimize for the worst case attacker solution. In this paper, we just deal with the case in which the inner utility has a unique global maximum.

## 3    Solution Method

Bilevel optimization problems can rarely be solved analytically. Indeed even extremely simple instances of bilevel problems have been shown to be NP-hard, [16]. Thus, numerical techniques are required. Several classical and evolutionary approaches have been proposed to solve (1), as reviewed by [26]. When the inner problem adheres to certain regularity conditions, it is possible to reduce the bilevel optimization problem to a single level one replacing the inner problem with its Karush-Kuhn-Tucker (KKT) conditions. Then, evolutionary techniques could be used to solve this single-level problem, thus making possible to relax the upper level requirements. As, in general, this single-level reduction is not feasible, several other approaches have been proposed, such as nested evolutionary algorithms or metamodeling-based methods. However, most of these approaches lack scalability: increasing the number of upper level variables produces an exponential increase on the number of lower level tasks required to be solved being thus impossible to apply these techniques to solve large scale bilevel problems as the ones appearing in the context of AML.

In [5] the authors face the problem of solving Stackelberg games in the AML context. However, they focus on a very particular type of game which can be reformulated as a quadratic program. In this paper, we provide more general procedures to solve Stackelberg games that are useful in the AML paradigm in which decision spaces are continuous and high dimensional. To this end, we focus on gradient ascent techniques to solve bilevel optimization problems.

Let us assume that for any $\alpha$ the solution of the inner problem is unique. This solution defines an implicit function $\beta^*(\alpha)$. Thus, problem (1) may be viewed solely in terms of the defender decisions $\alpha$. The underlying idea behind gradient ascent techniques is the following: given a defender decision $\alpha \in \mathbb{R}^n$ a direction along which the defender's utility increases while maintaining feasibility must be found, and then, we move $\alpha$ in that direction. Thus, the major issue of ascent methods is to find the gradient of $u_D(\alpha, \beta^*(\alpha))$. In [18] the authors provide a method to approximate such gradient that work for relatively large classical optimization problems but it is clearly insufficient to deal with the typical bilevel problems appearing in AML.

Recently, [7] proposed forward and reverse-based methods for computing the gradient of the validation error in certain hyperparamenter optimization problems that appear in Deep Learning. Structurally, hyperparameter optimization problems are similar to Stackelberg games. We adapt their methodology to this domain. In particular we propose two alternative approaches to compute the gradient of $u_D[\alpha, \beta^*(\alpha)]$ with different memory and running time requirements. We refer to these approaches as backward and forward solutions, respectively.

**Notation.** For the sake of clarity, we use the following notation: the gradient will be denoted as $d_x$; the partial derivative as $\partial_x$. Similarly, second partial derivatives will be denoted as $\partial_x^2$ and $\partial_x \partial_y$. We shall use this notation indistinctly for the unidimensional and multidimensional cases. For instance, if $f(x, y)$ is a scalar function, $x$ is a $p$-dimensional vector and $y$ is a $q$-dimensional vector, then $\partial_x^2 f(x, y)$ is the $p \times p$ matrix whose $(i, j)$ entry is $\partial_{x_i} \partial_{x_j} f(x, y)$, where $x_i$ is the $i$-th component of the vector $x$. Similarly, $\partial_x \partial_y f(x, y)$ is a $p \times q$ matrix whose $i, j$ entry is $\partial_{x_i} \partial_{y_j} f(x, y)$.

### 3.1    Backward Solution

We propose here a new gradient ascent approach to solve the bilevel problem (1) whose running time scales well with the defender's decision space dimension. In particular, we propose to approximate problem (1) by the following PDE-constrained optimization problem, [15]

$$\arg\max_{\alpha} \quad u_D\left[\alpha, \beta(\alpha, T)\right]$$

$$\text{s.t.} \qquad \partial_t \beta(\alpha, t) = \partial_\beta u_A[\alpha, \beta(\alpha, t)] \tag{2}$$

$$\beta(\alpha, 0) = 0.$$

The idea is formalized in the next proposition, that can be proved using the results in [3].

**Proposition 1.** *Suppose that the following assumptions hold*

1. *The attacker problem, the inner problem in* (1), *has a unique solution* $\beta^*(\alpha)$ *for each defender decision* $\alpha$.
2. *For all* $\epsilon > 0$ *and all* $\alpha$,

$$\inf_{\|\beta - \beta^*(\alpha)\|_2^2 > \epsilon} \langle \beta - \beta^*(\alpha), \partial_\beta u_A[\alpha, \beta] \rangle > 0.$$

*If* $\beta(\alpha, t)$ *satisfies the differential equation*

$$\partial_t \beta(\alpha, t) = \partial_\beta u_A[\alpha, \beta(\alpha, t)] \tag{3}$$

*then* $\beta(\alpha, t) \to \beta^*(\alpha)$ *as* $t \to \infty$, *with rate* $\mathcal{O}\left(\frac{1}{t}\right)$.

The idea in (2) is thus to constrain the trajectories $\beta(\alpha, t)$ to satisfy (3) and approximate the defender's problem using $\beta(\alpha, T)$ with $T \gg 1$, instead of $\beta^*(\alpha)$.

We propose solving problem (2) using gradient ascent and the adjoint method, [24], to compute the total derivative of the defender utility function with respect to her decision. The adjoint method defines an *adjoint function* $\lambda(t)$ satisfying the *adjoint equation*

$$d_t \lambda(t) = -\lambda(t) \, \partial_\beta^2 u_A[\alpha, \beta(\alpha, t)]. \tag{4}$$

In terms of the adjoint function, the derivative of the defender utility with respect to her decision would be written as

$$d_\alpha u_D[\alpha, \beta(\alpha, T)] = \partial_\alpha u_D[\alpha, \beta(\alpha, T)] - \int_0^T \lambda(t) \partial_\alpha \partial_\beta u_A[\alpha, \beta(\alpha, t)] \, dt. \tag{5}$$

In Appendix A, we prove that if $\lambda(t)$ satisfies the adjoint equation (4), the derivative of the defender utility can be written as in (5).

Algorithmically, we can proceed by discretizing (4) via Euler method, and approximate the derivative (5) discretizing the integral on the left hand side. This leads to Algorithm 1. Once we are able to compute this derivative, we can solve the defender's problem using gradient ascent.

Regarding its complexity, note that by basic facts of Automatic Differentiation (AD), [12], if $\tau(n, m)$ is the time required to evaluate $u_D(\alpha, \beta)$ and $u_A(\alpha, \beta)$, then computing derivatives of these functions requires time $\mathcal{O}(\tau(n, m))$. Thus the first for loop in Algorithm 1 requires time $\mathcal{O}(T\tau(n, m))$. In the second loop, we

**Algorithm 1.** Approximate total derivative of defender utility function with respect to her decision using the backward solution

1: **procedure** APPROXIMATE DERIVATIVE USING BACKWARD METHOD$(\alpha, T)$
2:     $\beta_0(\alpha) = 0$
3:     **for** $t = 1, 2, \ldots, T$ **do**
4:         $\beta_t(\alpha) = \beta_{t-1}(\alpha) + \eta \partial_\beta u_A(\alpha, \beta)\big|_{\beta_{t-1}}$
5:     **end for**
6:     $\lambda_T = -\partial_\beta u_D(\alpha, \beta)\big|_{\beta_T}$
7:     $\mathrm{d}_\alpha u_D = \partial_\alpha u_D[\alpha, \beta_T(\alpha)]$
8:     **for** $t = T - 1, T - 2, \ldots, 0$ **do**
9:         $\mathrm{d}_\alpha u_D = \mathrm{d}_\alpha u_D - \lambda_{t+1} \partial_\alpha \partial_\beta u_A(\alpha, \beta)\big|_{\beta_t}$
10:         $\lambda_t = \lambda_{t+1}\left[I - \partial_\beta^2 u_A(\alpha, \beta)\big|_{\beta_t}\right]$
11:     **end for**
12:     **return** $\mathrm{d}_\alpha u_D$
13: **end procedure**

need to compute second derivatives, which appear always multiplying the vector $\lambda_t$. By basic results of AD, Hessian vector products have the same time complexity as function evaluations. Thus in our case, we can compute second derivatives in time $\mathcal{O}(\tau(n, m))$ being the time complexity of the second for loop $\mathcal{O}(T\tau(n, m))$. Thus, overall, Algorithm 1 runs in time $\mathcal{O}(T\tau(n, m))$. Regarding space complexity, as it is necessary to store the values of $\beta_t(\alpha)$ produced in the first loop for later usage in the second one, if $\sigma(n, m)$ is the space requirement for storing each $\beta_t(\alpha)$, then $\mathcal{O}(T\sigma(n, m))$ is the space complexity of the backward algorithm.

In certain applications where space complexity is critical, the backward solution could be infeasible as, within each iteration, it requires storing the whole trace $\beta_t(\alpha)$. In this particular cases, the forward solution proposed in the next section, solves this issue at a cost of loosing time scalability.

### 3.2   Forward Solution

In this case, we approximate (1) by

$$\arg\max_{\alpha} \quad u_D[\alpha, \beta_T(\alpha)]$$

$$\text{s.t} \qquad \beta_t(\alpha) = \beta_{t-1}(\alpha) + \eta_t \partial_\beta u_A(\alpha, \beta)\big|_{\beta_{t-1}} \qquad t = 1, \ldots, T \qquad (6)$$

$$\beta_0(\alpha) = 0.$$

The idea here is that, for each defense $\alpha$, we condition on a dynamical system that under certain conditions converges to $\beta^*(\alpha)$, the optimal solution for the attacker when the defender plays $\alpha$. Thus, we can approximate the defender's utility by $u_D[\alpha, \beta_T(\alpha)]$, with $T \gg 1$. This idea is formalized in the next proposition that can be proved using the results of [3].

**Proposition 2.** *Suppose that the following assumptions hold*

1. *The attacker problem (the inner problem in (1)) has a unique solution $\beta^*(\alpha)$ for each defender decision $\alpha$.*
2. *For all $\epsilon > 0$ and $\alpha$*

$$\inf_{\|\beta - \beta^*(\alpha)\|_2^2 > \epsilon} \langle \beta - \beta^*(\alpha), \partial_\beta u_A[\alpha, \beta] \rangle > 0$$

3. *For some $A, B > 0$ and all $\alpha$*

$$\|\partial_\beta u_A[\alpha, \beta]\|_2^2 \leq A + B\|\beta - \beta^*(\alpha)\|_2^2$$

*If for all $t$, $\beta_t(\alpha)$ satisfies*

$$\beta_t(\alpha) = \beta_{t-1}(\alpha) + \eta \partial_\beta u_A(\alpha, \beta)\Big|_{\beta_{t-1}} \tag{7}$$

*Then, $\beta_t(\alpha)$ converges to $\beta^*(\alpha)$, with rate $\mathcal{O}\left(\frac{1}{t}\right)$.*

We propose solving problem (6) using gradient ascent. To that end, we need to compute $d_\alpha u_D(\alpha, \beta_T(\alpha))$. Using the chain rule we have

$$d_\alpha u_D[\alpha, \beta_T(\alpha)] = \partial_\alpha u_D[\alpha, \beta_T(\alpha)] + \partial_{\beta_T} u_D[\alpha, \beta_T(\alpha)] \, d_\alpha \beta_T(\alpha)$$

To obtain $d_\alpha \beta_T(\alpha)$, we can sequentially compute $d_\alpha \beta_t(\alpha)$ taking derivatives in (7)

$$d_\alpha \beta_t(\alpha) = d_\alpha \beta_{t-1}(\alpha) + \eta_{t-1} \left[ \partial_\alpha \partial_\beta u_A(\alpha, \beta)\Big|_{\beta_{t-1}} + \partial_\beta^2 u_A(\alpha, \beta)\Big|_{\beta_{t-1}} d_\alpha \beta_{t-1}(\alpha) \right]$$

This induces a dynamical system in $d_\alpha \beta_t(\alpha)$ that can be iterated in parallel to the dynamical system in $\beta_t(\alpha)$. The whole procedure is described in Algorithm 2. Once we are able to compute this derivative, we can solve the defender's problem using gradient ascent.

Regarding time complexity, note that the bottleneck in Algorithm 2 is that we need to compute second derivatives of $u_A(\alpha, \beta)$. In particular, computing $\partial_\beta^2 u_A(\alpha, \beta)$ requires time $\mathcal{O}(m\tau(m, n))$ as it requires computing $m$ Hessian vector products, one with each of the $m$ the unitary vectors. On the other hand, computing $\partial_\alpha \partial_\beta u_A(\alpha, \beta)$ requires computing $n$ Hessian vector products and thus time $\mathcal{O}(n\tau(m, n))$, while if we compute the derivative in the other way, first we derive with respect to $\beta$ and then with respect to $\alpha$, the time complexity is $\mathcal{O}(m\tau(m, n))$. Thus, we derive first with respect to the variable with the biggest dimension. Then, the time complexity of computing $\partial_\alpha \partial_\beta u_A(\alpha, \beta)$ is $\mathcal{O}(\min(n, m)\tau(m, n))$. Finally, as $\partial_\beta^2 u_A(\alpha, \beta)$ and $\partial_\alpha \partial_\beta u_A(\alpha, \beta)$ could be computed in parallel, then the overall time complexity of the forward solution is $\mathcal{O}(\max[\min(n, m), m]T\tau(m, n)) = \mathcal{O}(mT\tau(m, n))$. Regarding space, as in this case the values $\beta_t(\alpha)$ are overwritten at each iteration, we do not need to store all of them and the overall space complexity is $\mathcal{O}(\sigma(m, n))$.

**Algorithm 2.** Approximate total derivative of defender utility function with respect to her decision using the forward solution.

1: **procedure** APPROXIMATE DERIVATIVE USING FORWARD METHOD($\alpha, T$)
2:     $\beta_0(\alpha) = 0$
3:     $d_\alpha \beta_0(\alpha) = 0$
4:     **for** $t = 1, 2, \ldots, T$ **do**
5:         $\beta_t(\alpha) = \beta_{t-1}(\alpha) + \eta \partial_\beta u_A(\alpha, \beta)\Big|_{\beta_{t-1}}$
6:         $d_\alpha \beta_t(\alpha) = d_\alpha \beta_{t-1}(\alpha) + \eta_{t-1} \left[ \partial_\alpha \partial_\beta u_A(\alpha, \beta)\Big|_{\beta_{t-1}} + \partial_\beta^2 u_A(\alpha, \beta)\Big|_{\beta_{t-1}} d_\alpha \beta_{t-1}(\alpha) \right]$
7:     **end for**
8:     $d_\alpha u_D = \partial_\alpha u_D[\alpha, \beta_T(\alpha)] + \partial_{\beta_T} u_D[\alpha, \beta_T(\alpha)] d_\alpha \beta_T(\alpha)$
9:     **return** $d_\alpha u_D$
10: **end procedure**

## 4  Experiments

We illustrate now the proposed approaches. We start with a conceptual example in which we empirically test the scalability properties of both algorithms. Then, we apply the algorithms to solve a problem in the context of adversarial regression.

All the code used for these examples has been written in python using the pytorch library for Automatic Differentiation, [23]; and is available at https://github.com/roinaveiro/GM_SG.

### 4.1  Conceptual Example

We use a simple example to illustrate the scalability of the proposed approaches. Consider that the attacker's and defender's decisions are both vectors in $\mathbb{R}^n$. The attacker's utility takes the form $u_A(\alpha, \beta) = -\sum_{i=1}^n 3(\beta_i - \alpha_j)^2$ and the defender's one is $u_D(\alpha, \beta) = -\sum_{i=1}^n (7\alpha_i + \beta_j^2)$. In this case, the equilibrium can be computed analytically using backward induction: for a given defense $\alpha \in \mathbb{R}^n$ we see that $\beta^*(\alpha) = \alpha$; substituting in the outer problem, the equilibrium is reached at $\alpha_j^* = -3.5, \beta_j^*(\alpha^*) = -3.5$ with $j = 1, \ldots, m$.

We apply the proposed methods to this problem to test their scalability empirically. The parameters were chosen as follows: the learning rate $\eta$ of Algorithms 1 and 2 was set to 0.1; similarly, the learning rate of the gradient ascent used to solve the outer problem was also set to 0.1. Finally, all gradient ascents were run for $T = 40$, enough to reach convergence.

Figure 2 shows running times for increasing number of dimensions of the decision spaces (in this problem both the attacker's and the defender's decision space have the same dimension). As we discussed, the forward running time increases linearly with the number of dimensions while the backward solution remains approximately constant. This obviously comes at the cost of having more memory requirements, as in Algorithm 1 we need to store the whole trace $\beta_t(\alpha)$. Thus, in problems where the dimension of $\beta$ is very large the memory

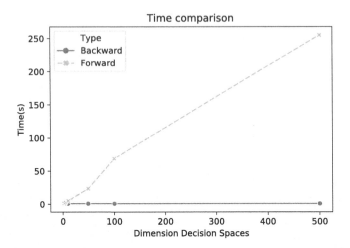

**Fig. 2.** Backward and Forward running times versus the dimension of decision spaces.

cost of the backward solution would become prohibitive and we would need to switch to the forward solution, as long as the dimension of $\alpha$ is small enough. In contrast, if the dimension of $\alpha$ is very big, the forward solution would become infeasible in time, thus being the backward optimal provided that the dimension of $\beta$ is such that it is possible to store the whole trace $\beta_t(\alpha)$.

### 4.2 An Application to Adversarial Regression

**Problem Statement.** We illustrate an application of the proposed methodology to adversarial regression problems, [13]. They are a specific class of prediction games, [5], played between a *learner* of a regression model and a *data generator*, who tries to fool the learner modifying input data at application time, inducing a change between the data distribution at training and test time, with the aim of confusing the data generator and attain a benefit.

Given a feature vector $x \in \mathbb{R}^p$ and its corresponding target value $y \in \mathbb{R}$, the learner's decision is to choose the weight vector $w \in \mathbb{R}^p$ of a linear model $f_w(x) = x^\top w$, that minimizes the theoretical costs at application time, given by

$$\theta_l(w, \bar{p}, c_l) = \int c_l(x, y)(f_w(x) - y)^2 \, \mathrm{d}\bar{p}(x, y),$$

where $c_l(x, y) \in \mathbb{R}^+$ reflects instance-specific costs and $\bar{p}(x, y)$ is the data distribution at test time. To do so, the learner has a training matrix $X \in \mathbb{R}^{n \times p}$ and a vector of target values $y \in \mathbb{R}^n$, that is a sample from distribution $p(x, y)$ at training time.

The data generator aims at changing features of test instances to induce a transformation in the data distribution from $p(x, y)$ to $\bar{p}(x, y)$. Let $z(x, y)$ be the data generator's target value for instance $x$ with real value $y$, i.e. he aims at transforming $x$ to make the learner predict $z(x,y)$ instead of $y$. The data generator aims at choosing the transformation that minimizes the theoretical costs given by

$$\theta_d(w, \bar{p}, c_d) = \int c_d(x,y)(f_w(x) - z(x,y))^2 \, d\bar{p}(x,y) + \Omega_d(p, \bar{p})$$

where $\Omega_d(p, \bar{p})$ is the incurred cost when transforming $p$ to $\bar{p}$ and $c_d(x,y)$ are instance specific costs.

As the theoretical costs defined above depend on the unknown distributions $p$ and $\bar{p}$, we focus on their regularized empirical counterparts, given by

$$\widehat{\theta}_l(w, \bar{X}, c_l) = \sum_{i=1}^{n} c_{l,i}(f_w(\bar{x}_i) - y_i)^2 + \Omega_l(f_w),$$

$$\widehat{\theta}_d(w, \bar{X}, c_d) = \sum_{i=1}^{n} c_{d,i}(f_w(\bar{x}_i) - z_i)^2 + \Omega_d(X, \bar{X}).$$

In addition, we assume that the learner acts first, choosing a weight vector $w$. Then the data generator, after observing $w$, chooses his optimal data transformation. Thus, the problem to be solved by the learner is

$$\arg\min_{w} \quad \widehat{\theta}_l(w, T(X, w, c_d), c_l)$$
$$\text{s.t.} \qquad T(X, w, c_d) = \arg\min_{X'} \widehat{\theta}_d(w, X', c_d), \tag{8}$$

where $T(X, w, c_d)$ is the attacker's optimal transformation for a given choice $w$ of weight vector. (8) has the same form as (1), except that it is formulated in terms of costs rather than utilities. In addition, it is easy to see that if $\Omega_d(X, \bar{X})$ is equal to the squared Frobenius norm of the difference matrix $\|X - \bar{X}\|_F^2$, then the attacker's problem has a unique solution. Thus, we can use the proposed solution techniques to look for Nash equilibria in this type of game, taking care of performing gradient descent instead of gradient ascent, as we are minimizing costs here.

**Experimental Results.** We apply the results to the UCI white wine dataset, [6]. This contains real information about 4898 wines, that consists of 11 quality indicators plus a wine quality score.

$R_J$ and $R_D$ are two competing wine brands. $R_D$ has implemented a system to automatically measure wine quality using a regression over the available quality indicators: each wine is described by a vector of 11 entries, one per quality indicator. Wine quality ranges between 0 and 10. $R_J$, aware of the actual superiority of its competitor's wines, decides to hack $R_D$'s system by manipulating the value of several quality indicators, to artificially decrease $R_D$'s quality rates. However, $R_D$ is aware of the possibility of being hacked, and decides to use adversarial methods to train its system. In particular, $R_D$ models the situation as a Stackelberg game. It is obvious that the target value of his enemy is $z(x, y) = 0$ for every possible wine. In addition, $R_D$ was able to filter some information about $R_J$'s wine-specific costs $c_{d,i}$.

As basic model, a regular ridge regression, [9], was trained using eleven principal components as features. The regularization strength was chosen using repeated hold-out validation, [17], with ten repetitions. As performance metric we used the root mean squared error (RMSE), estimated via repeated hold-out.

We compare the performance of two different learners against an adversary whose wine specific costs $c_{d,i}$ are fixed: The first one, referred to as Nash, assumes that the wine specific costs are common knowledge and plays Nash equilibria of the Stackelberg game defined in (8). The second learner, refer to as raw, is a non adversarial one and uses a ridge regression model. To this end, we split the data in two parts, 2/3 for training purposes and the remaining 1/3 for test. The training set is used to compute the weights $w$ of the regression problem. Those weights are observed by the adversary, and used to attack the test set. Then, the RMSE is computed using this attacked test set and the previously computed weights.

In order to solve (8), we use the backward solution method of Sect. 3.1 due to its better time scalability. The hyperparameters were chosen as follows: number of epochs $T$ to compute the gradient in Algorithm 1, 100; the learning rate $\eta$ in this same Algorithm, was set to 0.01. Within the gradient descent optimization used to optimize the defender's cost function, the number of epochs was set to 350 and the learning rate to $10^{-6}$. Finally, we assumed that the wine specific costs were the same for all instances and called the common value $c_d$. We studied how $c_d$ affects the RMSE for different solutions.

Notice that, in this case, the dimension of the attacker's decision space is huge. He has to modify the training data to minimize his costs. If there are $k$ instances in the training set, each of dimension $n$, the dimension of the attacker's decision space is $n \times k$. In this case $k = 3263$ (2/3 of 4898) and $n = 11$. Thus the forward solution is impractical in this case, and we did not compute it. We show in Fig. 3, the RMSE for different values of the wine specific cost. We observe that Nash outperforms systematically the adversary unaware regression method. In the limit $c_d \rightarrow 0$, we see that $\widehat{\theta}_d(w, \bar{X}, c_d) \rightarrow \Omega_d(X, \bar{X})$. Thus, in this situation, the adversary will not manipulate the data. Consequently Nash and ridge regression solutions will coincide, as shown in Fig. 3. However, as $c_d$ increases, data manipulation is bigger, and the RMSE of the adversary unaware method also increases. On the other hand, the Nash solution RMSE remains almost constant.

We have also computed the average and standard deviation of training times. In an Intel Core i7-3630UM, 2.40 GHz 8 computer, the average training time is 131.6 s with 2.7 s standard deviation. This corresponds approximately to 2.66 s per outer epoch. Each outer epoch involves running Algorithm 1 with 100 inner epochs.

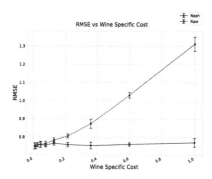

**Fig. 3.** Performance comparison.

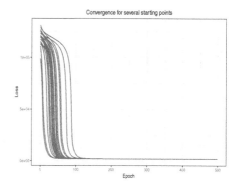

**Fig. 4.** Convergence for several initial points.

Finally, to illustrate convergence of the proposed approach, we solve (8) using gradient descent with the backward method for 20 different random initializations of the defender's decisions $\omega$. Results are depicted in Fig. 4. As can be seen, all paths converge with less than 150 epochs.

## 5  Discussion

The demand for scalable solutions of Stackelberg Games has increased in the last years due to the use of such games to model confrontations within Adversarial Machine Learning problems. In this paper, we have focused on gradient methods for solving Stackelberg Games, providing two different approaches to compute the gradient of the defender's utility function: the forward and backward solutions. In particular, we have shown that the backward solution scales well in time with the defender's decision space dimension, at a cost of more memory requirements. On the other hand, the forward solution scales poorly in time with this dimension, but well in space.

We have provided empirical support of the scalability properties of both approaches using a simple example. In addition, we have solved an AML problem using the backward solution in a reasonable amount of time. In this problem, the defender's decision space is continuous with dimension 11. The attacker's decision space is also continuous with dimension $\mathcal{O}(10^4)$, as we showed in Sect. 4.2. To the best of our knowledge, none of previous numerical techniques for solving Stackelberg games could deal, in reasonable time, with such high dimensional continuous decision spaces.

Apart from scalability properties, a major advantage of the proposed framework is that it could be directly implemented in any Automatic Differentiation library such as PyTorch (the one used in this example) or TensorFlow, and thus benefit from the computational advantages of such implementations.

We could extend the framework in several ways. First, as we discussed, the backward solution has poor space scalability. This is generally not an issue in most applications. Nevertheless, if space complexity is critical it is possible to reduce it at a cost of introducing a numerical error, as proposed in [19] in hyper-parameter optimization problems. Instead of storing the whole trace $\beta_t(\alpha)$ in the first for loop of Algorithm 1 to use it in the second loop, we could sequentially undo its gradient update at each step of the second for loop. Obviously, this would introduce some numerical error.

Another possible line of work would be to extend the framework to deal with Bayesian Stackelberg games, that are widely used to model situations in AML in which there is not common knowledge of the adversary's parameters. In this line, the ultimate goal would be to apply the proposed algorithms to solve Adversarial Risk Analyisis (ARA, [25]) problems in AML, [22].

Throughout the paper, we have focused on exact gradient methods. However, it would be interesting to extend the proposed algorithms to work with stochastic gradient methods. In addition, in [21] the authors propose several variants of Gradient Ascent to solve saddle point problems. It could be worth investigating how to extend such techniques to general Stackelberg Games.

Finally, we highlight that one of the most important contributions of the paper is the derivation of the backward solution formulating the Stackelberg game (1) as a PDE-constrained optimization problem and using the adjoint method. This provides a general and scalable framework that could be used to seek for Nash equilibria in other types of sequential games. Exploring this, is another possible line of future work.

## A    Proof of the Adjoint Method

The Lagrangian of problem (2) is

$$\mathcal{L} = u_D[\alpha, \beta(\alpha, T)] + \int_0^T \lambda(t) \left\{ \mathrm{d}_t \beta(\alpha, t) - \partial_\beta u_A[\alpha, \beta(\alpha, t)] \right\} \mathrm{d}t + \mu \beta(\alpha, 0).$$

As the constraints hold, by construction we have that $d_\alpha \mathcal{L} = d_\alpha u_D$ and

$$d_\alpha \mathcal{L} = \partial_\alpha u_D[\alpha, \beta(\alpha, T)] + \partial_\beta u_D[\alpha, \beta(\alpha, T)] \, d_\alpha \beta(\alpha, T) + \mu \, d_\alpha \beta(\alpha, 0) \qquad (9)$$
$$+ \int_0^T \lambda(t) \left\{ \mathrm{d}_t \, d_\alpha \beta(\alpha, t) - \partial_\alpha \partial_\beta u_A[\alpha, \beta(\alpha, t)] - \partial_\beta^2 u_A[\alpha, \beta(\alpha, t)] \, d_\alpha \beta(\alpha, t) \right\} \mathrm{d}t.$$

Integrating by parts, we have

$$\int_0^T \lambda(t) \, \mathrm{d}_t \, d_\alpha \beta(\alpha, t) \, \mathrm{d}t = \left[ \lambda(t) \, d_\alpha \beta(\alpha, t) \right]_0^T - \int_0^T \mathrm{d}_t \lambda(t) \, d_\alpha \beta(\alpha, t) \, \mathrm{d}t$$

Inserting this in (9) and grouping the terms conveniently we have

$$d_\alpha \mathcal{L} = \partial_\alpha u_D[\alpha, \beta(\alpha, T)] + \left\{ \partial_\beta u_D[\alpha, \beta(\alpha, T)] + \lambda(T) \right\} d_\alpha \beta(\alpha, T) + \left\{ \mu - \lambda(0) \right\} d_\alpha \beta(\alpha, 0)$$
$$+ \int_0^T \left\{ -\mathrm{d}_t \lambda(t) - \lambda(t) \partial_\beta^2 u_A[\alpha, \beta(\alpha, t)] \right\} d_\alpha \beta(\alpha, t) - \lambda(t) \partial_\alpha \partial_\beta u_A[\alpha, \beta(\alpha, t)] \, \mathrm{d}t$$

Since the constraints hold, we may choose freely the Lagrange multipliers. In particular, we may choose them so that we can avoid calculating the derivatives of $\beta(\alpha, t)$ with respect to $\alpha$ (as this is computationally expensive). Thus, we have that $\lambda$ satisfies the adjoint equation

$$\mathrm{d}_t \lambda(t) = -\lambda(t)\partial_\beta^2 u_A[\alpha, \beta(\alpha, t)]$$

with $\lambda(T) = -\partial_\beta u_D[\alpha, \beta(\alpha, T)]$, and $\mu = \lambda(0)$. Using this, the derivative is computed as

$$\mathrm{d}_\alpha \mathcal{L} = \partial_\alpha u_D[\alpha, \beta(\alpha, T)] - \int_0^T \lambda(t)\partial_\alpha\partial_\beta u_A[\alpha, \beta(\alpha, t)]\,\mathrm{d}t$$

# References

1. Bard, J.F.: Some properties of the bilevel programming problem. J. Optim. Theory Appl. **68**(2), 371–378 (1991)
2. Biggio, B., Roli, F.: Wild patterns: ten years after the rise of adversarial machine learning. Pattern Recogn. **84**, 317–331 (2018)
3. Bottou, L.: Online learning and stochastic approximations. On-Line Learn. Neural Netw. **17**(9), 142 (1998)
4. Brown, G., Carlyle, M., Salmerón, J., Wood, K.: Defending critical infrastructure. Interfaces **36**(6), 530–544 (2006)
5. Brückner, M., Scheffer, T.: Stackelberg games for adversarial prediction problems. In: Proceedings of the 17th ACM SIGKDD International Conference on Knowledge Discovery and Data Mining, pp. 547–555. ACM (2011)
6. Dheeru, D., Karra Taniskidou, E.: UCI machine learning repository (2017). http://archive.ics.uci.edu/ml
7. Franceschi, L., Donini, M., Frasconi, P., Pontil, M.: Forward and reverse gradient-based hyperparameter optimization. In: Proceedings of the 34th International Conference on Machine Learning, vol. 70. pp. 1165–1173. JMLR.org (2017)
8. French, S., Insua, D.R.: Statistical Decision Theory. Wiley, New York (2000)
9. Friedman, J., Hastie, T., Tibshirani, R.: The Elements of Statistical Learning. Springer Series in Statistics, vol. 1. Springer, New York (2001). https://doi.org/10.1007/978-0-387-21606-5
10. Gibbons, R.: A Primer in Game Theory. Harvester Wheatsheaf, New York (1992)
11. González-Ortega, J., Insua, D.R., Cano, J.: Adversarial risk analysis for bi-agent influence diagrams: an algorithmic approach. Eur. J. Oper. Res. **273**(3), 1085–1096 (2019)
12. Griewank, A., Walther, A.: Evaluating derivatives: principles and techniques of algorithmic differentiation, vol. 105 (2008)
13. Großhans, M., Sawade, C., Brückner, M., Scheffer, T.: Bayesian games for adversarial regression problems. In: International Conference on Machine Learning, pp. 55–63 (2013)
14. Heap, S.H., Varoufakis, Y.: Game Theory. Routledge, London (2004)
15. Hinze, M., Pinnau, R., Ulbrich, M., Ulbrich, S.: Optimization with PDE Constraints, vol. 23. Springer, Dordrecht (2008). https://doi.org/10.1007/978-1-4020-8839-1

16. Jeroslow, R.G.: The polynomial hierarchy and a simple model for competitive analysis. Math. Program. **32**(2), 146–164 (1985)
17. Kim, J.H.: Estimating classification error rate: repeated cross-validation, repeated hold-out and bootstrap. Comput. Stat. Data Anal. **53**(11), 3735–3745 (2009)
18. Kolstad, C.D., Lasdon, L.S.: Derivative evaluation and computational experience with large bilevel mathematical programs. J. Optim. Theory Appl. **65**(3), 485–499 (1990)
19. Maclaurin, D., Duvenaud, D., Adams, R.: Gradient-based hyperparameter optimization through reversible learning. In: International Conference on Machine Learning, pp. 2113–2122 (2015)
20. McDaniel, P., Papernot, N., Celik, Z.B.: Machine learning in adversarial settings. IEEE Secur. Priv. **14**(3), 68–72 (2016)
21. Mokhtari, A., Ozdaglar, A., Pattathil, S.: A unified analysis of extra-gradient and optimistic gradient methods for saddle point problems: Proximal point approach. arXiv preprint arXiv:1901.08511 (2019)
22. Naveiro, R., Redondo, A., Insua, D.R., Ruggeri, F.: Adversarial classification: an adversarial risk analysis approach. Int. J. Approx. Reason. **113**, 133–148 (2019)
23. Paszke, A., et al.: Automatic differentiation in PyTorch (2017)
24. Pontryagin, L.S.: Mathematical Theory of Optimal Processes. Routledge, London (2018)
25. Rios Insua, D., Rios, J., Banks, D.: Adversarial risk analysis. J. Am. Stat. Assoc. **104**(486), 841–854 (2009)
26. Sinha, A., Malo, P., Deb, K.: A review on bilevel optimization: from classical to evolutionary approaches and applications. IEEE Trans. Evol. Comput. **22**(2), 276–295 (2018)
27. Tambe, M.: Security and Game Theory: Algorithms, Deployed Systems, Lessons Learned. Cambridge University Press, Cambridge (2011)
28. Vorobeychik, Y., Kantarcioglu, M.: Adversarial machine learning. Synth. Lect. Artif. Intell. Mach. Learn. **12**(3), 1–169 (2018). https://doi.org/10.2200/S00861ED1V01Y201806AIM039

# Interactive Elicitation of a Majority Rule Sorting Model with Maximum Margin Optimization

Ons Nefla[1,2(✉)], Meltem Öztürk[1], Paolo Viappiani[3], and Imène Brigui-Chtioui[2]

[1] Université de Paris Dauphine, PSL Research University CNRS, LAMSADE,
75016 Paris, France
ons.nefla@dauphine.eu, meltem.ozturk@dauphine.fr
[2] Emlyon Business School, 23 Avenue Guy de Collongue, Ecully, France
brigui-chtioui@em-lyon.com
[3] Sorbonne Université, CNRS, Laboratoire d'informatique de Paris 6, LIP6,
75005 Paris, France
paolo.viappiani@lip6.fr

**Abstract.** We consider the problem of eliciting a model for ordered classification. In particular, we consider *Majority Rule Sorting (MR-sort)*, a popular model for multiple criteria decision analysis, based on pairwise comparisons between alternatives and idealized profiles representing the "limit" of each category.

Our interactive elicitation protocol asks, at each step, the decision maker to classify an alternative; these assignments are used as training set for learning the model. Since we wish to limit the cognitive burden of elicitation, we aim at asking informative questions in order to find a good approximation of the optimal classification in a limited number of elicitation steps. We propose efficient strategies for computing the next question and show how its computation can be formulated as a linear program. We present experimental results showing the effectiveness of our approach.

**Keywords:** Preference elicitation · Ordinal classification ·
Incremental elicitation · MR-sort · Simulations

## 1 Introduction

There are several situations where it is necessary to classify objects, defined on several criteria, into ordered classes (for example, credit ratings, evaluating students, hotel categorization, etc). Such ordinal classification problems, also called *multi-criteria sorting*, have been considered by ELECTRE TRI [16,19], a popular method from the field of multi-criteria decision analysis that has been successfully applied to several domains.

*Majority Rule Sorting (MR-sort)* [15] is a simplified version of ELECTRE TRI belonging to a class of "non compensatory" decision models that have been

© Springer Nature Switzerland AG 2019
S. Pekeč and K. B. Venable (Eds.): ADT 2019, LNAI 11834, pp. 141–157, 2019.
https://doi.org/10.1007/978-3-030-31489-7_10

axiomatized by Bouyssou and Marchant [5]. The remarkable characteristic of the MR-sort procedure is that its classifications are readily explainable to the users. In MR-sort an alternative is assigned to a category when it is "at least as good as" an idealized profile describing the category's lower "limit" and "not better" than the category's upper limit. The limit profiles encode the characteristics that an alternative should have in order to be assigned to a particular class according to the decision maker (for example: a 4-stars hotel should have rooms of at least a certain size, it should have a swimming pool, a 5 star hotel should have a classy restaurant, etc). The relation comparing alternatives and limit profiles is based on a weighted majority rule (in Sobrie et al. [21] a non linear model for combining the weights is considered).

In order to use MR-sort in practice, it is necessary to undertake a preference elicitation phase in order to assess the parameters of the model. The "classical" approach to elicitation consists in a fine grained assessment of the model parameters; however since it is often not reasonable to directly ask the decision maker about the values of the parameters of the model, we adopt an incremental elicitation approach and we assume that the decision maker can easily classify one given alternative into one of the categories; the information provided by the user can be used to update the model and provide a better result. The intuition is that, by selectively choosing which items the decision maker could classify, we could retrieve a good approximation of the "correct" model of the decision maker (DM) even with a small number of examples. We also stress that the information provided by the user may be noisy, as she may have made a mistake in assessing the right category for the alternative.

In this paper we adopt max-margin optimization, whose main principle is the following. The currently known assignments of items to categories are encoded by a set of inequalities on the feasible parameters; a shared non-negative margin is introduced as a decision variable that is maximized in the objective function. Noisy feedback is addressed by relaxing the constraints using slack variables and adding a penalty term in the objective function for violated constraints. In this paper, we propose an incremental approach for eliciting the parameters of a MR-sort model; at each step of the elicitation procedure, the system asks a question to the user and the question is specifically chosen in order to be as "informative" as possible. Our main algorithm repeatedly uses the max-margin optimization routine with two goals (1) to make an estimation of the model parameters given the current information (2) to determine a measure of uncertainty associated with a potential question. For the second goal, we consider different strategies based on computing the maximum-margin a posteriori (considering the different answers that the decision maker could give).

We now provide a brief discussion about related works. Reasoning with preferences [7] is an important issue in artificial intelligence. Several researchers have dealt with the problem of dealing with an incompletely specified preference model and with the issue of preference elicitation [4,6,24,27]. In the machine learning community, approaches for *preference learning* have been proposed [10], including approaches for ordinal classification (see Chapter 8 in [17], and, for instance [9] and [11]).

Our work is stimulated by recently proposed approaches for eliciting multi-attribute utility functions using maximum-margin optimization [25,26] in configuration problems. The maximum-margin optimization that we adopt has been used (with some variations) in previous works [21,22] that tackled the problem of learning the weights of a MR-sort model; these works, however, did not consider incremental elicitation. Bennabou et al. [2,3] recently proposed interactive elicitation methods based on minimax regret for ordinal classification problems (with a related but different model); these methods however have the inconvenience that they are not tolerant to errors in the responses of the decision maker.

The paper is structured as follows. After some background in Sect. 2, we present MR-sort model in Sect. 3 and the interactive elicitation in Sect. 4. We then discuss our experimental results in Sect. 5 and provide some final comments and directions for future work in Sect. 6.

## 2    Background

We consider a set $X$ of $m$ items that are evaluated with respect to a *set of criteria* $\mathscr{C} = \{criterion_1, \ldots, criterion_n\}$. Criteria associate items with a performance evaluation; (with little abuse of notation, we use $x$ to refer to both the item and its evaluation vector). The *evaluation of $x$* is a vector $(x_1, x_2, \ldots, x_n) \in \mathcal{E}_1 \times \ldots \mathcal{E}_n$. The sets $\mathcal{E}_1, \ldots, \mathcal{E}_n$ are totally ordered and represent the space of criteria evaluations. Indeed we are interested in ordinal classification methods that allow the criteria evaluations to be defined on scales not necessarily numerical, and that can differ among the criteria. We use $[m]$ to denote the set $\{1, \ldots, m\}$.

MR-sort [15] is a multi-criteria ordinal classification method allowing to assign alternatives to ordered categories. The *set of categories* is denoted by $\mathcal{C} = \{C_1, C_2, \ldots, C_p\}$. Categories are ordinal, $C_1$ being the worst and $C_p$ the best one. Each category $C_h$ is characterized by two "fictitious" items $b^h = (b_1^h, \ldots, b_n^h) \in \mathcal{E}_1 \times \ldots \mathcal{E}_n$ and $b^{h+1} = (b_1^{h+1}, \ldots, b_n^{h+1}) \in \mathcal{E}_1 \times \ldots \mathcal{E}_n$; these are called the *limit profiles of $C_h$* and we denote by $B = \{b^1, b^2 \ldots, b^{p+1}\}$ the set of such limit profiles. Limit profiles play the role of the lower and the upper bounds of the category $C_h$; limit profiles of higher categories dominate the lower ones: $\forall k = 1, \ldots, n$, $b_k^{i+1} \geq_k b_k^i$ and $\exists j, b_j^{i+1} >_k b_j^i$, where $\geq_k$ is the binary relation on the evaluations on the $criterion_k$. There are two special limit profiles, $b^1$ and $b^{p+1}$ that are defined as the minimum and the maximum values: $b^1 = (b_1^1, \ldots, b_n^1)$, such that $\forall i, b_i^1 = \min_{x \in X}(x_i)$ (resp. $b^{p+1} = (b_1^{p+1}, \ldots, b_n^{p+1})$, such that $b_i^{p+1} = \max_{x \in X}(x_i)$). To fully specify a MR-sort model we need to associate each criterion $cr_i$ with a numerical weight $w_i$, that intuitively represents its importance. A parameter $\lambda$ is called the *majority threshold*, whose role will become clear below.

The procedure for making assignments is based on pairwise comparisons between objects of $X$ and limit profiles. An alternative $x$ is assigned to the category $C_h$ if it is "at least as good as" the lower limit profile $b^h$ and it is not "at least as good as" the upper limit profile $b^{h+1}$ according to a binary relation $\succeq$. Indifference: $x \sim y \iff x \succeq y$ and $y \succeq x$.

$$x \to C_h \iff x \succeq b^h \text{ and } x \not\succeq b^{h+1} \tag{1}$$

where $x \to C_h$ means that alternative $x$ is assigned to category $C_h$. An item $x$ that is indifferent to the upper limit profile of the best category ($x \sim b^{p+1}$, meaning $x \succeq b^{p+1}$ and $x \preceq b^{p+1}$) is assigned to $C_p$. The binary relation $\succeq \subseteq (\mathcal{E}_1 \times \ldots \mathcal{E}_n) \times (\mathcal{E}_1 \times \ldots \mathcal{E}_n)$ is based on the *weighted majority principle*:

$$x \succeq b^h \iff \sum_{i: x_i \succeq_i b_i^h} w_i \geq \lambda \tag{2}$$

*Example.* Assume that we have 5 hotels $(x, y, z, t, u)$ defined on $n = 3$ criteria (cost, comfort, quality of the restaurant). Our aim is to assign our hotels into $p = 3$ categories ($C_1$: 1 star, $C_2$: 2 stars, $C_3$: 3 stars). The MR-sort model parameters are: $\lambda = 0.55$, $w = (0.2, 0.5, 0.3)$. The criteria evaluation scale is between 0 and 5 (5 being the best score) for $criterion_1$ and $criterion_2$; the scale for $criterion_3$ is between 0 and 10. Table 1 presents the performances of the limit profiles and the alternatives; Table 2 their comparisons using $\succeq$ and the assignments of alternatives.

**Table 1.** Performance of limit profiles $b^1, b^2, b^3, b^4$ and alternatives $x, y, z, t, u$

|       | $criterion_1$ | $criterion_2$ | $criterion_3$ |
|-------|-----|-----|-----|
| $b^1$ | 0   | 0   | 0   |
| $b^2$ | 2   | 2   | 4   |
| $b^3$ | 4   | 4   | 8   |
| $b^4$ | 5   | 5   | 10  |
| $x$   | 1   | 2   | 3   |
| $y$   | 3   | 3   | 8   |
| $z$   | 1   | 5   | 8   |
| $t$   | 1   | 3   | 10  |
| $u$   | 1   | 2   | 1   |

**Table 2.** Comparison between alternatives and profiles and final assignments.

|       | $b^1$ | $b^2$ | $b^3$ | $b^4$ | Assignment |
|-------|-------|-------|-------|-------|------------|
| $x$   | $\succeq$ | $\not\succeq$ | $\not\succeq$ | $\not\succeq$ | $C_1$ |
| $y$   | $\succeq$ | $\succeq$ | $\not\succeq$ | $\not\succeq$ | $C_2$ |
| $z$   | $\succeq$ | $\succeq$ | $\succeq$ | $\not\succeq$ | $C_3$ |
| $t$   | $\succeq$ | $\succeq$ | $\not\succeq$ | $\not\succeq$ | $C_2$ |
| $u$   | $\succeq$ | $\not\succeq$ | $\not\succeq$ | $\not\succeq$ | $C_1$ |

Note that the MR-sort method allows the use of heterogeneous scales since the only information being useful for the relation $\succeq$ is an ordinal one. Hence, a problem with only ordinal scales or with different types of scales (interval one, ratio one and ordinal one) can be handled without difficulty by an MR-sort model.

## 3   Learning a MR-Sort Model from Assignments

We assume that we are given some assignments of alternatives to categories. In this work we assume that limit profiles are given. In the following, we first introduce the notion of possible and necessary categories, and then we present a maximum-margin optimization for learning the parameters of a MR-sort model.

## 3.1   Possible Categories

Let $\mathcal{A}$ be the set of alternatives whose category is known and LS the "learning set" (pairs of alternatives and categories): LS $= \{(x, C_h), x \in \mathcal{A}, x \to C_h\}$. From LS, assuming an underlying MR-sort model, we can reason about the weights that are consistent with the current knowledge. According to Eqs. (1) and (2), an assignment of the type $x \xrightarrow{\text{DM}} C_h$ (made by the decision maker) corresponds to imposing the linear constraints:

$$\sum_{i:x_i \geq b_i^h} w_i \geq \lambda \tag{3}$$

$$\sum_{i:x_i \geq b_i^{h+1}} w_i < \lambda \tag{4}$$

Let $\Theta(\text{LS})$ be set of parameters that are compatible with the learning set LS, satisfying constraints (3) and (4) for all pairs in LS, and the requirements that the weights are non negative and normalized.

When dealing with partially specified preference models, it is typical to reason about possible and necessary preference information [13]. In our context, given an alternative and its performance evaluation we can reason about the parameters that are consistent with the current knowledge.

Now, given an item $x \notin \mathcal{A}$, we define the set of possible categories PC$(x; \text{LS})$ as the set of categories such that there is an instantiation of the parameters $\theta = \{w_1, \ldots, w_n, \lambda\}$ consistent with the assignment *given the previously known assignments* LS.

$$\text{PC}(x; \text{LS}) = \{C_i \in \mathcal{C} \mid \exists \theta \in \Theta(\text{LS}) : x \xrightarrow{\theta} C_i\}$$

where we write $x \xrightarrow{\theta} C_i$ to emphasize the dependency between the parameters $\theta$ and the assignment. If there is only one possible category, i.e. $|\text{PC}(x)| = 1$, then it means that the alternative has to be *necessarily* assigned to the only category in PC (assuming that the model is consistent with the learning set; in Sect. 3.2 we discuss how to handle inconsistencies). In practice it is possible that a partial knowledge about the parameters allows us to determine in which categories we have to place several alternatives. We anticipate that the concept of possible categories plays an important role in our elicitation strategies (see Sect. 4.2).

## 3.2   Maximum Margin Optimization

We now address the problem of learning the parameters $\theta = \{w_1, \ldots, w_n, \lambda\}$ of a MR-sort model (given a set of learning assignments) with a linear program. We assume that we are given as input the following values: the number of criteria $n$, the number of categories $p$, and the limit profiles $b^h$ for each $h \in [p]$. Our goal is to assess the weights $w$ and the majority threshold $\lambda$, that are decision variables for the optimization problem. In order to discriminate between different choices for the parameters, the first step is to introduce a shared margin $\mu$ as an additional decision variable. The first linear program (LP) assumes that the data in LS is perfectly consistent with the MR-sort model.

$$\mu^* = \max \quad \mu \tag{5}$$

$$\text{s.t.} \quad \sum_{i:x_i, \geq b_i^h} w_i \geq \lambda + \mu \qquad\qquad \forall (x, C_h) \in \text{LS} \tag{6}$$

$$\sum_{i:x_i \geq b_i^{h+1}} w_i \leq \lambda - \mu \qquad\qquad \forall (x, C_h) \in \text{LS} \tag{7}$$

$$\sum_{i=1}^{n} w_i = 1 \tag{8}$$

$$\mu \in \mathbb{R} \tag{9}$$

$$w_i \geq 0 \qquad\qquad\qquad\qquad \forall i \in [n] \tag{10}$$

Constraints (6) and (7) correspond to Eqs. (3) and (4) with the inclusion of the shared margin $\mu$. Note that later in Sect. 4.2 we will use $\mu^*(\text{LS})$ to denote the application of the LP above to find the value of the best compatible margin.

We observe that in general there may be occasional inconsistencies in user feedback, or simply the data may not be compatible with a MR-sort model. The "typical" way to handle this is to introduce slack variables (whose sum we aim at minimizing); this idea goes back to the UTA approach [14] and several later models [1,12]. We formalize the problem of finding the weights of a MR-sort model (given a set of learning assignments) with the following linear program.

$$m^* = \max \quad \mu - \alpha \sum_{(x, C_h) \in \text{LS}} \xi_{i,j} \tag{11}$$

$$\text{s.t.} \quad \sum_{i:x_i, \geq b_i^h} w_i + \xi_{x,h} \geq \lambda + \mu \qquad\qquad \forall (x, C_h) \in \text{LS} \tag{12}$$

$$\sum_{i:x_i \geq b_i^{h+1}} w_i - \xi_{x,h} \leq \lambda - \mu \qquad\qquad \forall (x, C_h) \in \text{LS} \tag{13}$$

$$\sum_{i=1}^{n} w_i = 1 \tag{14}$$

$$\mu \in \mathbb{R} \tag{15}$$

$$\xi_{x,h} \geq 0 \qquad\qquad\qquad\qquad \forall (x, C_h) \in \text{LS} \tag{16}$$

$$w_i \geq 0 \qquad\qquad\qquad\qquad \forall i \in [n] \tag{17}$$

The variables are the following: $\lambda \in [0, 1]$, $w_i \in [0, 1] \forall i \in [n]$, $\mu \in [0, 1]$ and $\xi_{x,h} \in [0, 1] \ \forall x \in \text{LS}$ such that $x$ assigned to $C_h$. We use a parameter $\alpha$ to express the "cost" of violating a constraint representing assignments from the learning set. The objective (Eq. 11) is to maximize the shared margin $\mu$ (that we want to maximize) minus the sum of the constraints violations $\xi_{i,j}$ (that we want to minimize). In order to be able to handle the inconsistency in the user's answers we add a slack variable $(\xi_{x,h})$ for each $x \in \mathcal{A}$ and for each $h \in \{1, \ldots, p\}$. Constraint 12 and constraint 13 handle the conditions related to the assignments of the alternatives in the learning set LS. Constraint 14 enforces that the weights are normalized.

# 4   Incremental MR-sort

---

**Algorithm 1.** The MR-sort-Inc algorithm. K is the number of questions, X the set of alternatives, C the set of categories, LS $(LS = \{(x, C_h), x \in \mathcal{A}, x \rightarrow C_h\})$ learning assignments, s the strategy.

---

 1: Procedure MR-sort-Inc(K,X,C,LS,s)
 2: **for** $j \leftarrow 1, \ldots, K$ **do**
 3:     Compute $m^*(LS)$ and associated $w^*$ and $\lambda^*$ with linear program
 4:     Classify items in $X \setminus \mathcal{A}$ using $w^*$ and $\lambda^*$
 5:     Compute classification error          ▷ This is only possible in simulations
 6:     **for** $x \in X \setminus \mathcal{A}$ **do**          ▷ Evaluate items to decide next question
 7:         **for** $C_i \in C$ **do**
 8:             LS$' \leftarrow$ LS $\cup \{x \rightarrow C_i\}$          ▷ Include additional assignment
 9:             $v_i(x) \leftarrow m^*(\text{LS}')$          ▷ Compute margin *a posteriori*
10:         **end for**
11:         $S_s(x) \leftarrow$ aggregate $v(x)$ according to strategy $s$
12:     **end for**
13:     $x^* \leftarrow \arg\max_x(S_s(x))$          ▷ Decide what to ask next
14:     $C_{x^*} \leftarrow \text{Answer}(x^* \rightarrow ?)$          ▷ Ask question to DM
15:     LS $\leftarrow$ LS $\cup \{(x^*, C_{x^*})\}$
16: **end for**
17: **return** $w^*, \lambda^*$

---

In this section, we provide an elicitation method for an ordinal classification problem assuming that the preferences can be modeled by MR-sort.

## 4.1   Main Framework

Our elicitation procedure starts with a small learning set LS to which we add one-by-one new assignment examples. We ask questions of the type *"In which category should x be assigned to?"*; it is crucial to select "informative" items to ask about, in order to quickly converge (in few interaction cycles) to a good classification model.

We remind that a MR-sort model is defined by the following parameters: the weights $(w_1, \ldots, w_n)$, the limit profiles $(b^1, \ldots, b^{p+1})$ and majority threshold $\lambda$. We fix the limit profiles before the beginning of the elicitation. In the experiments below the limit profiles are chosen in a way to evenly partition the criteria scale.

The main steps of our approach are presented in Algorithm 1. Briefly, we start with some data from the learning set LS (learning alternatives $\mathcal{A}$ and their classification) and we include them as constraints in our linear program (LP). Given the learning set, we can estimate the weights $w^*$ and $\lambda^*$ using the linear program; in simulations we can also evaluate the classification error. After that, we test the assignment of each unassigned item to each category and retrieve the margin values obtained when using the optimization routine with each of the additional assignments. This gives us a vector of margins "a posteriori", that are aggregated differently depending on the "strategy" (different strategies will be discussed in the following section), giving a score for each item. The item having the highest score is chosen: the user is asked about the assignment of that item and the learning set is augmented. This procedure continues until a stopping condition (in our experiments when we reach a fixed number of questions, but in real applications termination may be left to the user). The last linear program contains all the assignments of the computed learning set, hence it provides the weights that we are looking for.

*Example.* We apply our incremental elicitation algorithm to the running example. Suppose that we only know the assignment of alternative $x$ (LS $= \{(x, C_1)\}$) and we want to ask just one question to the user. We first set the limit profiles: $b^1 = (0, 0, 0), b^2 = (\frac{5}{3}, \frac{5}{3}, \frac{10}{3}), b^3 = (\frac{10}{3}, \frac{10}{3}, \frac{20}{3}), b^4 = (5, 5, 10)$. Our goal is to learn $\lambda$ and a weight vector $w'$ which will assign the remaining alternatives to categories as close as possible to the ones presented on page 4. For this, according to our algorithm, we compute the score of alternatives $y, z, t, u$ $(S_s(y), S_s(z), S_s(t), S_s(u))$ and ask the assignment of the alternative having the highest score. We add this new constraint to LS and find the weight vector $w^*$ corresponding to the largest margin. Using $w^*$, we can find the current best assignment of the remaining alternatives and compute[1] an error measure using the true assignments given on page 4 and the one that we find using $w^*$.

## 4.2   Question Selection Strategies

In the following we present several strategies to select the next question to ask to the user. These strategies are used within our interactive elicitation paradigm described in Algorithm 1. The proposed strategies makes use of the max-margin optimization programs (discussed above in Subsect. 3.2) to identify informative questions.

**Most Uncertain.** With this strategy, we aim to ask the DM to classify the *most uncertain* item in $X \backslash \mathcal{A}$, that is the item that is compatible (according to the constraints derived from the known assignments of the learning set) with the highest number of categories. The score that this strategy assigns to an alternative $x$ is given by the cardinality of the set of possible categories: $S_{MU}(x) = |PC(x; LS)|$.

---

[1] This step is of course only to be performed in simulations, in real use of the procedure the classification error will not be known.

The computation of the set $PC(x)$ for an item $x$ makes use of the linear program for learning the parameters of a MR-sort without the slack variables (Eqs. 5–10). For each category $C_i$, we then constrain the item $x$ to be assigned to $C_i$ and simply check whether the margin is non negative; indeed $\mu^*(\text{LS} \cup (x, C_i)) > 0$ if and only if $C_i$ is a possible category ($C_i \in PC(x; \text{LS})$). The score of this strategy can be compactly written as:

$$S_{MU}(x) = |PC(x; \text{LS})| = \sum_{i=1}^{p} H[\mu^*(\text{LS} \cup (x, C_i))]$$

where $H(\cdot)$ is the step function.

*Example.* Consider the problem presented in the running Example. We suppose that we know the performance of each alternative, and the assignment of $x$, while the other assignments are unknown. Assume that we want to ask just one question. As before, we fix $b'^1 = (0, 0, 0), b'^2 = (\frac{5}{3}, \frac{5}{3}, \frac{10}{3}), b'^3 = (\frac{10}{3}, \frac{10}{3}, \frac{20}{3}), b'^4 = (5, 5, 10)$. Table 3 presents the $S_{MU}$ of all the remaining alternatives (step 3).

**Table 3.** Number of possible categories for the example of hotel categorization.

| Alternatives | $PC(x; \text{LS})$ | $S_{MU}$ |
|---|---|---|
| $y$ | $\{C_2, C_3\}$ | 2 |
| $z$ | $\{C_1, C_3\}$ | 2 |
| $t$ | $\{C_1, C_2, C_3\}$ | 3 |
| $u$ | $\{C_1\}$ | 1 |

As a result we ask the assignment of alternative $t$ to the user. The user will answer $C_2$ to this question (coherent with Example 2) and the constraint related to $t \rightarrow C_2$ will be added to the LP. After the inclusion of the constraint $t \rightarrow C_2$ in LS, we find our final weight vector $w^*$, such as $w^* = (0.266, 0.366, 0.366)$. Even if $w'$ is different from the weight vector of Example 2 ($w = (0.2, 0.5, 0.3)$), after our incremental elicitation we find the same assignments for $y, z, u$ (i.e. $y \rightarrow C_2, z \rightarrow C_3$ and $u \rightarrow C_1$).

**Sum-of-Margin Strategy.** The problem of the most-uncertain strategy is that, roughly speaking, it is agnostic to whether a potential assignment is consistent with large portions of the parameter space or just with small area. The intuition of sum-of-margin is to use the value of the objective function of the LP as a surrogate measure of the "degree" of satisfaction of an assignment $x \rightarrow C_h$. Intuitively, we should ask about items that may fit well into several different categories. We also include the possibility that some constraints can be violated, therefore make use of the second LP (Eqs. 11–17), with the penalty variables $\xi$ for violated assignments of the learning set.

Considering an alternative $x \in X \setminus \mathcal{A}$, we evaluate the penalized margin $m^*$ adding the assignment $x \to C_i$ for all $i \in [p]$ and construct the following vector:

$$v(x) = (m^*(\mathrm{LS} \cup (x, C_1)), \ldots, m^*(\mathrm{LS} \cup (x, C_p))).$$

In order to aggregate this vector into a single numerical measure, we adopt the sum. Hence, the score of the alternative $x$ is computed as

$$S_\Sigma(x) = \sum_{i=1}^{p} v_i(x) = \sum_{i=1}^{p} m^*(\mathrm{LS} \cup (x, C_i)).$$

**Table 4.** The score obtained by the different items in the hotel categorization example using the heuristics sum-of-maring (left) and entropy (right)

| Assignment $v_i(\cdot)$ | | $S_\Sigma$ | Assignment | $v'$ | $S_E$ |
|---|---|---|---|---|---|
| $y \to C_1$ | -0.45 | | $y \to C_1$ | 0 | |
| $y \to C_2$ | 0.45 | 0.45 | $y \to C_2$ | 0.5 | 0.30 |
| $y \to C_3$ | 0.45 | | $y \to C_3$ | 0.5 | |
| $z \to C_1$ | 0.45 | | $z \to C_1$ | 0.5 | |
| $z \to C_2$ | 0 | 0.90 | $z \to C_2$ | 0 | 0.30 |
| $z \to C_3$ | 0.45 | | $z \to C_3$ | 0.5 | |
| $t \to C_1$ | 0.45 | | $t \to C_1$ | 0.43 | |
| $t \to C_2$ | 0.18 | 1.08 | $t \to C_2$ | 0.15 | 0.44 |
| $t \to C_3$ | 0.45 | | $t \to C_3$ | 0.42 | |
| $u \to C_1$ | 0.49 | | $u \to C_1$ | 1 | |
| $u \to C_2$ | 0 | -0.10 | $u \to C_2$ | 0 | 0.00 |
| $u \to C_3$ | -0.55 | | $u \to C_3$ | 0 | |

*Example.* Consider again the running example. Table 4 (in the left) presents the $S_\Sigma$ of alternatives $\{y, z, t, u\}$. As the example shows, the best question is to ask about $t$, the second best question is to ask about $z$, then $y$, and finally $u$. We note a disagreement of MU and sum-of-margin about the ranking of $z$ and $y$.

**Entropy.** This strategy adopts the notion of entropy to assess the uncertainty for a given alternative. As the strategy sum-of-margin, we calculate $m^*$ which represents the maximum penalized margin if we assign the alternative $x$ to the category $C_i$, for all $i \in [p]$. We then combine these values in an evaluation vector $v'$ that filters out negative values (assigning them a zero value) and we normalize so that the values sum up to one:

$$v_i'(x) = \frac{R(m^*(\mathrm{LS} \cup (x, C_i)))}{\sum_{j=1}^{p} R(m^*(\mathrm{LS} \cup (x, C_j)))}$$

where $R(\cdot)$ is the ramp function. We then apply the entropy to calculate the final score of the alternative.

$$S_E(x) = - \sum_{i \in [p]: v_i'(x) > 0} v_i'(x) \log(v_i'(x))$$

Intuitively this method should favour asking about items whose uncertainty is the greatest in an information-theoretic sense, considering the vector $v'$ (composed of normalized non-negative margin values) as a surrogate measure for the probability distribution of the category of $x$ (i.e. how likely is $x$ to be in each of the categories).

*Example.* We consider again our example on hotel categorization. We now apply the entropy strategy; Table 4 (on the right) presents the score $S_E$ of all the alternatives. Item $t$ has the highest score, then $z$ and $y$ are tied in the second position; $u$ is last.

**Random Strategy.** As a baseline, we consider the random selection of an object $x$ from $X \setminus \mathcal{A}$. Utilizing this strategy is equivalent to use a non incremental elicitation procedure with $|LS| + K$ ($LS$ being the initial learning set of Algorithm 1 and $K$ being the number of questions) alternatives in the learning set.

Note that all our strategies have the same computational complexity: $O(Kmp)$ (resp. number of questions, alternatives, categories).

## 5   Experiments

We implemented our incremental MR-sort algorithm using Java API of CPLEX for solving the LP.[2] We performed simulations aimed at evaluating the effectiveness of the proposed elicitation strategies with two data-sets: a synthetic data-set fully represented by a MR-sort model, and a data-set from the UCI machine learning repository [8].

*Synthetic Data-Set.* For the first part of our experiments, the input data was generated randomly using a uniform distribution on the space of evaluations.

We include only one assignment, chosen at random in LS. The steps of the simulation are as follows:

i. Using a uniform distribution for each of the parameters, we generate the performance table of alternatives, a vector of weight $w$, the majority threshold $\lambda$. The limit profiles are chosen to evenly partition the range of evaluation values. These parameters fully specify the decision maker; we call this model $\overline{M}$. We then apply the MR-sort method in order to find the assignments of the generated alternatives; these assignments constitute the "ground truth".

---

[2] All experiments were run on a 2.9 GHz Intel, Core i7 and 16 Giga of RAM.

ii. We apply our incremental elicitation Algorithm 1 using the assignments of some alternatives of $\overline{M}$ as learning assignments. We apply incremental MR-sort by asking questions to the user based on the proposed strategies, simulating a decision maker who answers according to the results of the $\overline{M}$. We generate as output a vector of weight $w^*$ representing the DM's preferences.

iii. Using the assumed limit profiles and the learned weight vector $w*$ and the majority threshold $\lambda^*$, we obtain our learned model $M$ that provides the assignments of the remaining alternatives.

iv. At the end we calculate an error rate based on the difference between the assignments of the true model $\overline{M}$ and the learned model $M$.

We iterate these steps 50 times and we evaluate the classifications obtained with the different methods according to the average classification error $AE_k = \sum_{x \in X} err_k(x)/m$ where $m$ is the number of alternatives and $k \in \{1,2\}$. The value $err_k(x) = d(C_i, C_j)$ is the magnitude of error when $x \xrightarrow{\overline{M}} C_i$ and $x \xrightarrow{M} C_j$ (the true model assigns $x$ to $C_i$ while the learned model $M$ assigns it to $C_j$). $err_1$ adopts a 0/1 loss as distance, while $err_2$ considers the displacement between the assignments, that is $d(C_i, C_j) = |i-j|$.

Figure 1 presents the results of our simulations (for more results see Figs. 4, 5 and 6 in the appendix). Note that the random strategy performs very poorly, while all three strategies (maximum uncertainty, sum of margin, entropy) have reasonably good performance. Not surprisingly, the higher the number of categories, the quicker we converge, since there is more uncertainty with more categories. Conversely, the higher the number of criteria the less quickly the interaction converges. The results with respect to $AE_2$ are quite satisfying since they show that, after few questions have been answered, our procedure makes few assignment mistakes and these mistakes concern consecutive categories most of the times.

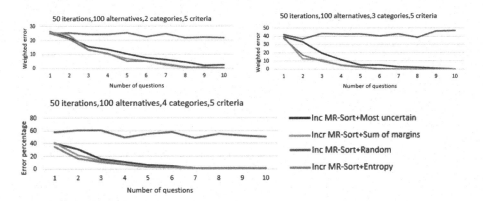

**Fig. 1.** $AE_2$ based on number of questions with $m = 100$, $n = 5$, $p = 2$, $p = 3$, $p = 4$, $|LS| = 1$ and 50 iterations

*UCI Car.* We performed simulations based on Car data-set. There are 1728 items defined on 6 categorical attributes (that we interpret as ordinal); the items are partitioned in 4 categories: unacceptable (65 items), acceptable (1210), good (384) and very good (69). We converted the qualitative (ordinal) attributes to numeric attributes and we set the criteria rating scale between 1 and 4. The steps of the simulation are: (i). We first compute the best error rate that we could have if all the data was included in $LS$. (ii). We ask questions until we reach this error percentage. We started to test on 300 alternatives chosen randomly and we start by including in the learning set only one assignment taken randomly.

In order to find the best error percentage, we call it $P^*$, we include all the data in LS and we vary the value of $\lambda$. We obtain $P^*$ when $\lambda$ is between 0.7 and 0.75. In this case the $err_1 = 18.86\%$ ($err_2$ being 0.24). So we aim to reach this percentage by asking the minimum number of questions. Our simulations showed that fixing the value of $\lambda$ makes faster the convergence, we show in Figs. 2, 3 the comparison between the 4 strategies where $\lambda$ is fixed to 0.7.

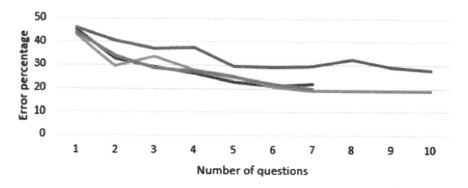

**Fig. 2.** $AE_1$ based on number of questions with $m = 300$, $n = 6$, $p = 4$, $|LS| = 1$ and 50 iterations

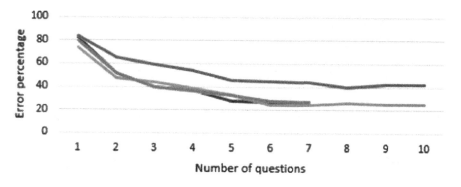

**Fig. 3.** $AE_2$ based on number of questions with $m = 300$, $n = 6$, $p = 4$, $|LS| = 1$ and 50 iterations

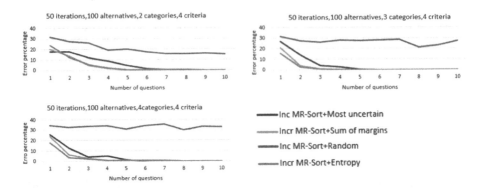

**Fig. 4.** Classification error $AE_1$ based on number of questions with $m = 100$, $n = 4$, $p = 2$, $p = 3$, $p = 4$, $|LS| = 1$ and 50 iterations

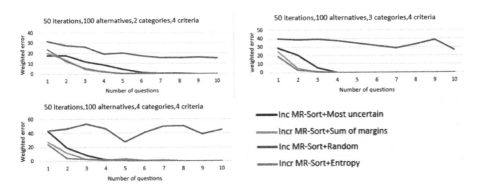

**Fig. 5.** $AE_2$ based on number of questions with $m = 100$, $n = 4$, $p = 2$, $p = 3$, $p = 4$, $|LS| = 1$ and 50 iterations

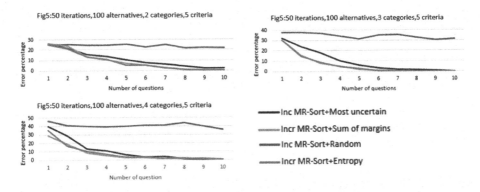

**Fig. 6.** $AE_1$ based on number of questions with $m = 100$, $n = 5$, $p = 2$, $p = 3$, $p = 4$, $|LS| = 1$ and 50 iterations

With the random strategy we observe that the variation of error according to number of questions is not monotonous and we don't approach to $P^*$ by asking 10 questions. On the other hand, the strategies maximum uncertainty and sum of margin are closer to $P^*$ starting from $6^{th}$ question. The strategy maximum uncertainty stops after the $7^{th}$ question because it hangs in case of inconsistency. We notice that the assignments errors are almost always between consecutive categories; Fig. 3 shows the performance with respect to the second metric.

# 6   Conclusions

MR-sort is an attractive method for ordinal classification that offers the advantage of allowing the use of heterogeneous scales (mixing ordinal and cardinal scales). In this paper we presented an incremental elicitation procedure for the parameters (weights and threshold) of MR-sort. Our approach relies on a maximum-margin optimization that aims at satisfy as well as possible the known assignments, following approaches proposed in the literature for non-interactive learning of MR-sort and variations [15,21,22]. The novelty of this paper consists in showing how the optimization can be used to evaluate the uncertainty associated to the items whose category is not known yet; the choice of the next question is based on evaluating the margin a posteriori (the value of maximum-margin optimization when adding a possible response to a question). Based on this intuition we proposed several strategies for selecting the next question to ask. We evaluated the proposed interactive elicitation procedure comparing the performance of the different strategies showing that the procedure quickly converges to the real optimal classification in very few interaction cycles.

We highlight that one important limitation of our framework is that we expect the decision maker to provide us with the limit profiles of the different categories. In future works we will relax this assumption considering techniques to elicit as well the limit profiles (either in a preliminary step or in an integrated approach), therefore providing a complete method for interactive elicitation of a MR-sort model. This task will be challenging, since previous works on (non incremental) elicitation of MR-sort have shown that, while it is possible to introduce integer variables [15] the resulting optimization is computationally very demanding and not scalable; therefore randomized heuristics [20,23] have been proposed.

We are also interested in performing simulation to compare our approach to other elicitation frameworks, as the recent work of Olteanu [18], and the approaches based on minimax regret [2,3], in realistic settings. We plan to investigate the connection between our approach and the field of machine learning. First, maximum-margin bears much similarity to Support Vector Machines (SVM). Second, ordinal classification has also been studied in machine learning. Third, there is strong similarity between incremental elicitation and active learning.

Finally, another important direction is to elicit the preferences of several users, providing methods that can exploit the similarity between users, as in Teso et al. [26].

# References

1. Ah-Pine, J., Mayag, B., Rolland, A.: Identification of a 2-additive bi-capacity by using mathematical programming. In: Perny, P., Pirlot, M., Tsoukiàs, A. (eds.) ADT 2013. LNCS (LNAI), vol. 8176, pp. 15–29. Springer, Heidelberg (2013). https://doi.org/10.1007/978-3-642-41575-3_2
2. Benabbou, N., Perny, P., Viappiani, P.: A regret-based preference elicitation approach for sorting with multicriteria reference profiles. In: From Multicriteria Decision Making to Preference Learning (DA2PL 2016) (2016)
3. Benabbou, N., Perny, P., Viappiani, P.: Incremental elicitation of choquet capacities for multicriteria choice, ranking and sorting problems. Artif. Intell. **246**, 152–180 (2017)
4. Boutilier, C.: Computational decision support: regret-based models for optimization and preference elicitation. In: Crowley, P.H. Zentall, T.R. (eds.) Comparative Decision Making: Analysis and Support Across Disciplines and Applications, pp. 423–453 (2013)
5. Bouyssou, D., Marchant, T.: An axiomatic approach to noncompensatory sorting methods in MCDM, II: more than two categories. Eur. J. Oper. Res. **178**(1), 246–276 (2007)
6. Conitzer, V., Sandholm, T.: Vote elicitation: complexity and strategy-proofness. In: AAAI/IAAI, pp. 392–397 (2002)
7. Domshlak, C., Hüllermeier, E., Kaci, S., Prade, H.: Preferences in AI: an overview (2011)
8. Dua, D., Graff, C.: UCI machine learning repository (2017)
9. Frank, E., Hall, M.: A simple approach to ordinal classification. In: De Raedt, L., Flach, P. (eds.) ECML 2001. LNCS (LNAI), vol. 2167, pp. 145–156. Springer, Heidelberg (2001). https://doi.org/10.1007/3-540-44795-4_13
10. Fürnkranz, J., Hüllermeier, E.: Preference Learning, 1st edn. Springer, Boston (2010). https://doi.org/10.1007/978-0-387-30164-8
11. Fürnkranz, J., Hüllermeier, E., Mencía, E.L., Brinker, K.: Multilabel classification via calibrated label ranking. Mach. Learn. **73**(2), 133–153 (2008)
12. Gajos, K., Weld, D.S.: Preference elicitation for interface optimization. In: Proceedings of UIST, pp. 173–182. ACM (2005)
13. Greco, S., Mousseau, V., Slowinski, R.: Ordinal regression revisited: multiple criteria ranking using a set of additive value functions. Eur. J. Oper. Res. **191**(2), 416–436 (2008)
14. Jacquet-Lagrèze, E., Siskos, Y.: Assessing a set of additive utility functions for multicriteria decision making: the UTA method. Eur. J. Oper. Res. **10**, 151–164 (1982)
15. Leroy, A., Mousseau, V., Pirlot, M.: Learning the parameters of a multiple criteria sorting method. In: Brafman, R.I., Roberts, F.S., Tsoukiàs, A. (eds.) ADT 2011. LNCS (LNAI), vol. 6992, pp. 219–233. Springer, Heidelberg (2011). https://doi.org/10.1007/978-3-642-24873-3_17
16. Mousseau, V., Slowinski, R., Zielniewicz, P.: A user-oriented implementation of the ELECTRE-TRI method integrating preference elicitation support. Comput. OR **27**(7–8), 757–777 (2000)

17. Murphy, K.P.: Machine Learning: A Probabilistic Perspective. The MIT Press, Cambridge (2012)
18. Olteanu, A.L.: Strategies for the incremental inference of majority-rule sorting models. In: DA2PL 2018: from Multiple Criteria Decision Aid to Preference Learning, Poznan, Poland, November 2018
19. Roy, B.: The outranking approach and the foundations of electre methods. In: Bana e Costa, C.A. (ed.) Readings in Multiple Criteria Decision Aid, pp. 155–183. Springer, Heidelberg (1990). https://doi.org/10.1007/978-3-642-75935-2_8
20. Sobrie, O., Mousseau, V., Pirlot, M.: Learning a majority rule model from large sets of assignment examples. In: Perny, P., Pirlot, M., Tsoukiàs, A. (eds.) ADT 2013. LNCS (LNAI), vol. 8176, pp. 336–350. Springer, Heidelberg (2013). https://doi.org/10.1007/978-3-642-41575-3_26
21. Sobrie, O., Mousseau, V., Pirlot, M.: Learning the parameters of a non compensatory sorting model. In: Walsh, T. (ed.) ADT 2015. LNCS (LNAI), vol. 9346, pp. 153–170. Springer, Cham (2015). https://doi.org/10.1007/978-3-319-23114-3_10
22. Sobrie, O., Mousseau, V., Pirlot, M.: A population-based algorithm for learning a majority rule sorting model with coalitional veto. In: Trautmann, H., et al. (eds.) EMO 2017. LNCS, vol. 10173, pp. 575–589. Springer, Cham (2017). https://doi.org/10.1007/978-3-319-54157-0_39
23. Sobrie, O., Mousseau, V., Pirlot, M.: Learning monotone preferences using a majority rule sorting model. Int. Trans. Oper. Res. **26**(5), 1786–1809 (2019)
24. Soufiani, H.A., Parkes, D.C., Xia, L.: Preference elicitation for general random utility models. arXiv preprint arXiv:1309.6864 (2013)
25. Teso, S., Passerini, A., Viappiani, P.: Constructive preference elicitation by setwise max-margin learning. In: Proceedings of the Twenty-Fifth International Joint Conference on Artificial Intelligence, IJCAI 2016, New York, NY, USA, 9–15 July 2016, pp. 2067–2073 (2016)
26. Teso, S., Passerini, A., Viappiani, P.: Constructive preference elicitation for multiple users with setwise max-margin. In: Rothe, J. (ed.) ADT 2017. LNCS (LNAI), vol. 10576, pp. 3–17. Springer, Cham (2017). https://doi.org/10.1007/978-3-319-67504-6_1
27. Zhao, Z., et al.: A cost-effective framework for preference elicitation and aggregation. arXiv preprint arXiv:1805.05287 (2018)

# Short Papers

# Approximating the Pareto Front of Bi-Criteria Kidney Exchanges

William Bailey$^{(\boxtimes)}$, Judy Goldsmith, Brent Harrison, and Siyao Xu

University of Kentucky, Lexington, KY 40506, USA
wba234@g.uky.edu, goldsmit@cs.uky.edu,
{brent.harrison,johnny.xu}@uky.edu

**Keywords:** Organ exchange · Bi-criteria optimization · Approximate pareto sets

Indirect organ exchange programs match pairs of incompatible donors and recipients with other pairs. The problem is typically phrased as a graph, for which we must find a maximal clearing; i.e., a set of disjoint cycles maximizing the number of organ recipients. Many maximal matchings exist, leaving the problem open to secondary optimization according to other criteria, including biological predictors of success [2], social issues [6], and logistic concerns [1]. To provide decision support for multi-criteria optimization where humans must remain in the loop, e.g., due to ethical concerns, we propose to provide decision-makers with the full range of Pareto optimal solutions. Because Pareto fronts of multi-criteria optimization problems may be exponential in the size of the instance, we focus on providing a representative subset, in the form of an $\epsilon$-Pareto set, or $\epsilon$-cover.

**Definition 1.** ([3] $\epsilon$-**Pareto set**). *Let $\mathcal{I}$ be an instance of $N$-criteria minimization. An $\epsilon$-Pareto set $P_\epsilon$ of $\mathcal{I}$ is a set of feasible solutions to $\mathcal{I}$ that approximately dominate every other feasible solution to $\mathcal{I}$. I.e, for each $p = (p_1, \ldots, p_N)$, there exists $c = (c_1, \ldots, c_N)$ in $P_\epsilon$ such that for all $i \in [N]$, $c_i \leq (1 + \epsilon)p_i$. When this inequality holds, we say that $p$ is $\epsilon$-covered, or $\epsilon$-dominated, by $c$.*

Computing minimal $\epsilon$-covers is *NP*-hard [3]. Therefore, we seek an approximation to the minimal set. In the general, two-objective case, the best PTAS provides a 3-approximation [5], although some special cases admit a 2-approximation [3].

© Springer Nature Switzerland AG 2019
S. Pekeč and K. B. Venable (Eds.): ADT 2019, LNAI 11834, pp. 161–163, 2019.
https://doi.org/10.1007/978-3-030-31489-7

**Algorithm 1.** Computes an $\epsilon$-cover $S$ of bi-criteria optimization instance $\mathcal{I}$, and an upper bound $\epsilon_{\max}$ on the MCE of $\mathcal{I}$.

$x_{\min} = \text{OPTIMIZE}(0, \mathcal{I})$
$y_{\min} = \text{OPTIMIZE}(1, \mathcal{I})$
$S = \{x_{\min}, y_{\min}\}, \epsilon_{\max} = 0$
$Q.\text{enqueue}(\text{TR}(x_{\min}, y_{\min}))$
**while** $Q \neq \{\}$ **and**
    $\text{UNEX-MCE}(Q.top()) > \epsilon_{\max}$ **do**
    $t = Q.\text{dequeue}()$
    $s = \text{OPTIMIZE}(t, \mathcal{I})$
    **if** $s \notin S$ **then**
        $S = S \cup \{s\}$
        $Q.\text{enqueue}(\text{TR}(\text{LEFT}(t, S), s))$
        $Q.\text{enqueue}(\text{TR}(s, \text{RIGHT}(t, S)))$
    **else**
        $\epsilon_{\max} = \max\{\epsilon_{max}, \text{EX-MCE}(t)\}$
    **end if**
**end while**
**return** $S, \epsilon_{\max}$

We characterize another general procedure (Algorithm 1) for finding $\epsilon$-covers in the two-objective case. Algorithm 1 uses a series of linear trade-offs between the criteria to approximate the Pareto front, returning an $\epsilon$-Pareto set $S$, and an upper bound $\epsilon_{\max}$ on the $\epsilon$ value provided by $S$. When compared to minimal $\epsilon$-covers providing this upper bound, simulated organ exchanges indicate that the algorithm is typically less than a factor of three of the minimum for $\epsilon_{\max}$. Algorithm 1 uses several subroutines. $\text{OPTIMIZE}(t, \mathcal{I})$ returns the maximal matching minimizing $C_t = (1 - t)x_1 + tx_2$, where $x_1$ and $x_2$ are the values of the optimization criteria. $TR(s_1, s_2)$ returns the value of $t$ such that $C_t(s_1) = C_t(s_2)$. LEFT$(t)$ and RIGHT$(t)$ return the nearest points in $S$ to the left and right, respectively, of the solution related to $t$.[1]

EX-MCE and UNEX-MCE relate to the stopping condition and the order in which trade-off values are used by OPTIMIZE. Each accepts two adjacent points from $S$: $s_1, s_1$. EX-MCE returns the minimum $\epsilon$ value for which all possible Pareto-optimal solutions between $s_1$ and $s_2$ would be covered. This upper bound can be improved if there are no convex solutions between $s_1$ and $s_2$, EX-MCE returns this improved bound. Together these *minimum covering epsilon* values provide an upper bound for the true $\epsilon$ value of the cover provided by Algorithm 1.[2]

The notion of using an $\epsilon$-cover is in line with recent work by McElfresh *et al.* [4] that calls for a division of labor for organ exchanges, involving policymakers, who decide on, and are held accountable for, the objectives pursued in situations like kidney exchange; and technicians, who provide recommendations based on the chosen objectives, but who are required to remain unbiased, and maintain *informed neutrality*. Our experiments show that $\epsilon$-covers may be a feasible, informed neutrality-preserving alternative for technicians who must simplify the choice for policymakers.

---

[1] Note that every solution $s$ in $S$ is related to the value of $t$ which discovered it; $s$ being optimal for $C_t$.

[2] It's important to note that Algorithm 1 can only produce an $\epsilon$-cover made of convex solutions, and cannot produce a cover for arbitrarily small $\epsilon$.

# References

1. Abraham, D.J., Blum, A., Sandholm, T.: Clearing algorithms for barter exchange markets: Enabling nationwide kidney exchanges. In: Proceedings of the 8th ACM conference on Electronic Commerce, pp. 295–304. ACM (2007)
2. Chen, J.H., Hughes, P., Woodroffe, C., Ferrari, P.: Pre-and post-donation kidney function in donors of a kidney paired donation with unique criteria for donor glomerular filtration rate-a longitudinal cohort analysis. Transpl. Int. **32**(3), 291–299 (2018)
3. Diakonikolas, I., Yannakakis, M.: Small approximate Pareto sets for biobjective shortest paths and other problems. SIAM J. Comput. **39**(4), 1340–1371 (2009)
4. McElfresh, D., Conitzer, V., Dickerson, J.: Ethics and mechanism design in kidney exchange (2018), Working draft
5. Vassilvitskii, S., Yannakakis, M.: Efficiently computing succinct trade-off curves. Theoret. Comput. Sci. **348**(2–3), 334–356 (2005)
6. Waterman, A.D., Rodrigue, J.R., Purnell, T.S., Ladin, K., Boulware, L.E.: Addressing racial and ethnic disparities in live donor kidney transplantation: priorities for research and intervention. Semin. Nephrol. **30**, 90–98 (2010)

# Formal Property-Oriented Design of Voting Rules Using Composable Modules

Karsten Diekhoff, Michael Kirsten$^{(\boxtimes)}$ ![ORCID], and Jonas Krämer

Karlsruhe Institute of Technology (KIT), Karlsruhe, Germany
kirsten@kit.edu, {karsten.diekhoff,jonas.kraemer}@student.kit.edu

**Abstract.** Voting rules aggregate multiple individual preferences in order to make a collective decision. Commonly, these mechanisms are expected to respect a multitude of different notions of fairness and reliability, which must be carefully balanced to avoid inconsistencies. We present an approach for the sound and flexible design of voting rules from composable modules. Formal composition rules guarantee social choice properties from properties of the individual components. The approach can be applied to many voting rules from the literature.

**Keywords:** Social choice · Formal correctness · Modular design

## 1 Introduction

In an election, voters cast ballots to express individual preferences about eligible alternatives. From these preferences, a collective decision, i.e., a set of elected alternatives, is determined using a *voting rule*. Voting rules are commonly designed to meet various expectations for fairness and reliability, but no one general rule caters for every requirement, and every rule shows paradoxical behavior for some situation [1]. The *axiomatic method* permits the analysis of desired behavior by comparing and characterizing voting rules via rigorous guarantees in the form of formal properties. Designing voting rules towards such properties is generally challenging as their trade-off is inherently difficult and error-prone.

**Contribution.** We present an approach for the systematic and formal design of voting rules from compact composable modules with formal properties guaranteed by construction. This work gives the core component type and compositional structures, e.g., for sequential, parallel and loop composition, and illustrates how composition rules formally establish common social choice properties.

## 2 Property-Oriented Composition of Voting Rules

**Electoral Modules.** The foundation of our approach are *electoral modules*, a generalization of voting rules. Voting rules elect a set of alternatives from a

© Springer Nature Switzerland AG 2019
S. Pekeč and K. B. Venable (Eds.): ADT 2019, LNAI 11834, pp. 164–166, 2019.
https://doi.org/10.1007/978-3-030-31489-7

profile, i.e., a sequence of ranked ballots, and a nonempty set of alternatives $A$. Electoral modules are more general as they do not need to make final decisions, but instead partition $A$ into *elected, rejected* and *deferred* alternatives. Hence, if an electoral module always produces a nonempty set of elected alternatives $A_{elected}$, it directly induces a voting rule which elects $A_{elected}$.

**Compositional Structures.** Our approach's core structures are *sequential, parallel* and *loop composition*, as well as the *revision* of decisions by prior modules. When composing two electoral modules $m \triangleright n$ sequentially, the second module $n$ only decides on alternatives which $m$ defers and cannot reduce the alternatives already elected or rejected. A parallel composition $m\|_a n$ delegates the two set-triples of $m$ and $n$ to an *aggregator* $a$, another component type which combines two such triples into one triple. Moreover, we may revise choices from prior modules and defer them for further decisions using a revision structure $\downarrow$. Finally, a loop composition $m \circlearrowleft_t$ reiterates a module $m$ sequentially until either $m$'s iteration reaches a fixed point, or a *termination condition* $t$ holds, i.e., a component type which is simply a predicate on a triple of sets of alternatives.

**A Simple Example.** The well-known *Baldwin's rule* [2] can be sequentially composed with a loop structure of a module eliminating the alternative with the lowest Borda score and terminating when only one alternative remains, and a module which elects all deferred alternatives. This construction directly establishes, e.g., the Pareto property and Condorcet consistency as the loop may never reject a Condorcet winner and always rejects Pareto-dominated alternatives.

# 3    Related Work, Conclusion and Outlook

**Related Work.** Our electoral modules are based on less-formal components for hierarchical electoral systems from [4]. Other work designs voting rules less modularly for statistically guaranteeing social choice properties by machine learning [7]. Prior modular approaches target verification [5] or declarative combinations of voting rules [3], but ignore social choice properties. Specific compositional structures as presented in [6] are readily expressible by our structures.

**Conclusion.** Our approach enables flexible and intuitive compositions of voting rules from a small number of structures with precise and general interfaces, easily extended with further modules. This allows to formally establish common social choice properties from given component properties by rigorous composition rules.

**Outlook.** A formally verified application of our approach is underway.

# References

1. Arrow, K.J.: Social Choice and Individual Values. Yale University, 3rd edn. (2012)
2. Baldwin, J.M.: The technique of the Nanson preferential majority system of election. Royal Society of Victoria **39**, (1926)

3. Charwat, G., Pfandler, A.: Democratix: A Declarative Approach to Winner Determination. In: Walsh, T. (ed.) ADT 2015. LNCS (LNAI), vol. 9346, pp. 253–269. Springer, Cham (2015). https://doi.org/10.1007/978-3-319-23114-3_16
4. Grilli di Cortona, P.: Evaluation & Optimization of Electoral Systems. SIAM (1999)
5. Ghale, M.K., Goré, R., Pattinson, D., Tiwari, M.: Modular Formalisation and Verification of STV Algorithms. In: Krimmer, R., Volkamer, M., Cortier, V., Goré, R., Hapsara, M., Serdült, U., Duenas-Cid, D. (eds.) E-Vote-ID 2018. LNCS, vol. 11143, pp. 51–66. Springer, Cham (2018). https://doi.org/10.1007/978-3-030-00419-4_4
6. Narodytska, N.: Combining voting rules together. In: ECAI, FAIA, vol. 242. IOS (2012)
7. Xia, L.: Designing social choice mechanisms using machine learning. In: AAMAS. IFAAMAS (2013)

# The Complexity of Elections
# with Rational Actors

Piotr Faliszewski[1] and Marija Slavkovik[2(✉)]

[1] AGH University of Science and Technology, Kraków, Poland
faliszew@agh.edu.pl
[2] University of Bergen, Bergen, Norway
marija.slavkovik@uib.no

**Abstract.** Voting and elections are among the most common methods of making collective decisions. The voters express their preferences regarding the candidates, and a voting rule aggregates them to provide the final election winner. We believe that to better understand elections, it is important to consider also the rationales behind the voters' individual preferences when aggregating them. To do this, we need to model and execute elections in such a way that the "rationality" is not optional or subjective. In this paper we propose to extend the traditional election model with information about the reasons for voters' choices.

## 1 The Model

We first briefly recall the approval-based election model and then extend it. An approval election $E = (C, V)$ consists of a set of candidates $C = \{c_1, \ldots, c_m\}$ and a collection of voters $V = (v_1, \ldots, v_n)$. Each voter $v_i$ has an approval set $A_i$ that consists of those candidates from $C$ that $v_i$ approves of. The approval score of candidate $c_j$, denoted $\mathrm{score}_E(c_j)$, is defined as the number of voters that approve $c_j$. Formally, $\mathrm{score}_E(c_j) = |\{v_i \in V \mid c_j \in A_i\}|$. The set of approval winners, denoted $\mathcal{R}(E)$, consists of those candidates that receive the highest approval score in a given election. Typically, we expect to have only a single winner, but we have to take into account the possibility of ties. In practice, tie-breaking mechanisms are used when this happens, but in this paper we disregard this issue.

In an *active candidate* model, we assume that the candidates take the action of announcing the issues that they intend to address when in the office, and the voters judge if these agendas are sufficiently convincing for them to grant their approvals.

We are given a set of candidates $C = \{c_1, \ldots, c_m\}$, a set of voters $V = (v_1, \ldots, v_n)$, and a set $\mathcal{P}$ of political positions. Each candidate $c_i$ is associated with a position $p(c_i) \in \mathcal{P}$, and each voter $v_i$ has an evaluation function $f_i \colon \mathcal{P} \to \{True, False\}$ that specifies if the candidate approves a given position or not. The set $A_i$ of the candidates approved by voter $v_i$ is:

$$A_i = \{c_j \mid f_i(p(c_j)) = True\}.$$

© Springer Nature Switzerland AG 2019
S. Pekeč and K. B. Venable (Eds.): ADT 2019, LNAI 11834, pp. 167–169, 2019.
https://doi.org/10.1007/978-3-030-31489-7

This model allows us to formulate variants of the classic POSSIBLE WINNER problems, see *e.g.*, the works of [Konczak and Lang, 2005; Xia and Conitzer, 2011], but perhaps in a somewhat more realistic format.

**Definition 1.** *In the* POSSIBLE WINNER WITH ACTIVE CANDIDATES *(PWAC) problem, we are given an election* $E = (C, V, \mathcal{P})$, *where* $C$ *is a set of candidates,* $V$ *is a collection of voters (with their functions for evaluating positions), and* $\mathcal{P}$ *is a set of possible positions; the input also contains the preferred candidate* $c_p$. *Each candidate* $c \in C$ *is associated with a set* $P_c$ *of positions that he or she may assume. We ask if it is possible to associate each candidate* $c \in C$ *with a position from* $P_c$, *so that* $c_p$ *is a winner of the resulting election.*

The active candidate model is appealing because it seems to be capturing the natural dynamics present in political elections: The candidates announce the platforms on which they run, and each voter individually evaluates each of them.

*Single-Peaked Elections.* Consider an election where taxation level is the main issue. The set of possible positions of the candidates is $\mathcal{P} = [0, 1]$. Each candidate $c_j$ announces his or her ideal taxation level $p(c_j) \in [0, 1]$. Each voter also has his or her interval $[a_i, b_i] \subseteq [0, 1]$ of acceptable taxation levels. Voter $v_i$ approves candidate $c_j$ if $p(c_j) \in [a_i, b_i]$. Formally, each evaluation function $f_i$ is defined as follows (for each $x \in [0, 1]$):

$$f_i(x) = \begin{cases} True, & \text{if } x \in [a_i, b_i], \\ False, & \text{otherwise.} \end{cases}$$

We choose to model the sets of possible candidate positions in the PWAC problem so that for each candidate $c$, $P_c$ is an interval $[x_c, y_c]$. In this case, our problem is polynomial-time computable.

**Theorem 1.** *For the active candidate model, the* PWAC *problem is in* P.

*Proof.* For the possible winner problem it suffices to choose the position of the preferred candidate so that it is approved by as many candidates as possible, and the positions of the remaining candidates to be approved by as few voters as possible. If in consequence $c_p$ has at least as high approval score as every other candidate, then we accept. Otherwise we reject. Computing positions of the candidates is easy: Indeed, it boils down to finding a point from a given interval that intersect either as many as possible or as few as possible given intervals; this is a classic problem that can be solved by a simple greedy algorithm.

## 2   Summary

In addition to active candidates, we also intend to propose an *active voter model*, in which it is the voters that are represented with a set of issues they care about. A voter then approves of a candidate if that candidate satisfies (most) of the issues the voter cares about. In addition to the possible winners, we can also study necessary winners. It is also natural to consider both rational election models and explore further issues of strategic behavior.

**Acknowledgement.** This work was supported in part by the AGH university and the "Doktorat Wdrożeniowy" program of the Polish Ministry of Science and Higher Education.

# References

Konczak, K., Lang, J.: Voting procedures with incomplete preferences. In: Proceedings of the Multidisciplinary IJCAI-05 Worshop on Advances in Preference Handling, pp. 124–129 (2005)

Xia, L., Conitzer, V.: Determining possible and necessary winners given partial orders. J. Artif. Intel. Res. **41**, 25–67 (2011)

# An Approach to Approximating Dominance in CP-nets

Michael Huelsman$^{(\boxtimes)}$ and Mirosław Truszczyński

University of Kentucky, Lexington, UK
{michael.huelsman,mirek}@uky.edu

**Abstract.** One of the most extensively studied preference representations is conditional preference networks (CP-nets). For even some simple types of CP-nets the problem of determining dominance, which of two alternatives is better, is NP-hard. In order to overcome this difficulty we propose a method of approximating CP-nets which is inspired by another preference representation, lexicographic preferences. We show that CP-nets dominance can be approximated in polynomial time and present several results on the accuracy of the approximation.

**Keywords:** CP-nets · Lexicographic preference · Dominance approximation

## 1 Introduction

In real world decision making problems, agents may have complex preferences over large sets of alternatives. In order to reason over such domains efficiently, an agent's preferences need to be represented compactly. One method of preference representation is through conditional preference networks (CP-nets) [2]. CP-nets have a number of useful properties both of practical and theoretical interest. However, some preference reasoning tasks for CP-nets are computationally hard. In particular, the task to determine the preference relation between two alternatives, commonly referred to as the *dominance* problem, is known to be NP-hard for some rather simple cases of CP-nets [2] and PSPACE-complete in general [3]. In order to handle this intractability we present a preference relation which approximates the CP-net and can be computed in polynomial time.

Our approach approximates dominance for an acyclic CP-net in polynomial time by exploiting the implicit attribute importance information contained in a CP-net's dependency graph along with the CP-net's conditional preference tables. We extract this importance information in order to build a representation which resembles a lexicographic preference model. A similar approach has been used by Ahmed and Mouhoub [1] in order to help solve constrained optimization problems with CP-nets. Our definition of dominance using this importance information is inspired by work from Yaman, Walsh, Littman, and desJardins [4]. Their paper dealt with, among other concerns, lexicographic preference models which had equally important attributes.

This work was funded by the NSF under the grant number IIS-1618783.

S. Pekeč and K. B. Venable (Eds.): ADT 2019, LNAI 11834, pp. 170–171, 2019.
https://doi.org/10.1007/978-3-030-31489-7

## 2    Approach

Our approach centers around the generation of importance rankings which are consistent with a CP-net's dependency graph $G = (V, E)$. An importance ranking is consistent if $r(a) < r(b)$ if $(a, b) \in E$. Combining an importance ranking with the CP-net's CPTs allows us to define the following preference relation $\succ_{r,T}$:

**Definition 1.** Given a CP-net $G = (V, E)$, a consistent ranking $r$, and CPTs $T$ from $G$, $\alpha \succ_{r,T} \beta$ if for some attribute $a \in V$ we have $\alpha[a] \succ_T \beta[a]$, for all $b \in V$ such that $r(b) \leq r(a)$ we have $\alpha[b] \succeq_T \beta[b]$, and for all $c \in V$ such that $r(c) < r(a)$ we have $\alpha[c] = \beta[c]$.

This preference relation has two important properties. The first is that for two alternatives $\alpha$ and $\beta$ if $\alpha \succ \beta$ according to the original CP-net then $\alpha \succ_{r,T} \beta$, for any consistent $r$. Secondly, if the two alternatives are incomparable according to $\succ_{r,T}$ then they must be incomparable according to the original CP-net.

When the preference relation $\succ$ of a CP-net is approximated by $\succ_{r,T}$ errors may arise only when two alternatives are comparable by $\succ_{r,T}$, but are incomparable according to the original CP-net. We can also show that if there is no consistent $r$ such that $\alpha$ and $\beta$ are incomparable then all consistent rankings agree as to which alternative is preferred. This means that if we can consider all consistent rankings then we can build a more accurate approximation than that given by any single consistent ranking. This aggregation can be in computed in polynomial time with respect to the size of the CP-net by building a consistent ranking which finds two given alternatives incomparable. This approach does have the restriction that the CP-net used must be acyclic, as there is no consistent ranking for a cyclic CP-net. A key difference between this approach and some other heuristics/approximations is the ability to show incomparability and thus potentially better reflect the original CP-net.

## References

1. Ahmed, S., Mouhoub, M.: Constrained optimization with preferentially ordered outcomes. In: 2018 IEEE 30th International Conference on Tools with Artificial Intelligence (ICTAI), pp. 307–314. IEEE (2018)
2. Boutilier, C., Brafman, R.I., Domshlak, C., Hoos, H.H., Poole, D.: CP-nets: a tool for representing and reasoning with conditional ceteris paribus preference statements. J. Artif. Intell. Res. **21**, 135–191 (2004)
3. Goldsmith, J., Lang, J., Truszczynski, M., Wilson, N.: The computational complexity of dominance and consistency in CP-nets. J. Artif. Intell. Res. **33**, 403–432 (2008)
4. Yaman, F., Walsh, T.J., Littman, M.L., Desjardins, M.: Democratic approximation of lexicographic preference models. In: Proceedings of the 25th International Conference on Machine Learning, pp. 1200–1207. ACM (2008)

# Aggregation over Metric Spaces: Proposing and Voting in Elections, Budgeting, and Legislation

Gal Shahaf[1], Ehud Shapiro[1], and Nimrod Talmon[2($\boxtimes$)]

[1] Weizmann Institute of Science, Rehovot, Israel
{gal.shahaf,ehud.shapiro}@weizmann.ac.il
[2] Ben-Gurion University, Beersheba, Israel

**Motivation.** A thriving e-democracy [8] will have to address the choice of officers, committees, parameters (e.g., interest rate), budgets, and legislation. Today, social choice theory addresses each of these settings independently, offering different elicitation and aggregation methods for each, thus making the practical realization of an e-democracy untenable. In addition, current theory focuses on the act of voting, practically ignoring the need for an egalitarian process for determining which alternatives to vote upon. *Here, we present a unifying framework for all these social choice settings with a uniform elicitation and aggregation method, which is egalitarian in that voters cast both proposals and votes.*

Inspired by the spatial model [3], facility location [4], and the use of metrics and distances in social choice (e.g.,[1,2,5,6,9]), we view each social choice setting as a metric space. Votes as well as proposals represent ideal points of the voters and the distance from them induces rankings over alternatives. We explore Condorcet aggregation and a continuum of solution concepts, ranging from minimizing the sum of distances to minimizing the maximum distance.

A voting rule must be simple to be acceptable, hence we do not expect these sophisticated aggregation methods to determine the final decision. Rather, they may aid a deliberative process by articulating the tentative joint will of the voters, inspiring voters' change of mind (ideal points) as well as the formation of coalitions behind this or other ideal points. Ultimately, we expect an ideal point to be elected only if it has explicit majority support. We hope to employ the results presented here as a uniform foundation for e-democracy across the entire spectrum of social choice settings it would face.

**Formal Model.** Our model consists of a metric space $(X, d)$, where $X$ is a set of elements and $d : X \times X \to \mathbb{R}$ is a metric function. We assume $n$ voters, where voter $i$ provides her ideal element $v_i \in X$; we infer $i$'s ranking as follows: Voter $i$ prefers $x$ to $y$ whenever $d(v_i, x) < d(v_i, y)$. An *aggregation method* $\mathcal{R}$ gets a set of $n$ votes $V$ over a metric space $(X, d)$ and returns a point $\mathcal{R} \in X$ as the winner. A Condorcet winner is an element $x \in X$ that is not majority beaten. For $1 \leq p < \infty$, we define the $L_p$ estimator of $V$: $L_p(V) := argmin_{x \in X} \sum_{i \in [n]} d(v_i, x)^p$, and define $L_\infty(V) := argmin_{x \in X} \max_i \{d(x, v_i)\}$. To guarantee uniqueness, we

A full version is available[7]. We thank the Braginsky Center.

© Springer Nature Switzerland AG 2019
S. Pekeč and K. B. Venable (Eds.): ADT 2019, LNAI 11834, pp. 172–174, 2019.
https://doi.org/10.1007/978-3-030-31489-7

define: $\widetilde{L}_p(V) := \lim_{q \to p} L_q(V)$; notice that, as $\sum_{i \in [n]} d(v_i, x)^p$ is continuous, $\widetilde{L}_p(V) \subseteq L_p(V)$. We consider axiomatic properties: (1) an aggregation method $\mathcal{R}$ is *majoritarian* if $|\{v_i \in V : v_i = w\}| \geq |V|/2$ implies that $w \in \mathcal{R}(V)$. (2) $\mathcal{R}$ is *monotone* if for each $V = \{v_1, \ldots, v_{n-1}, v\}$ and a co-winner $w \in \mathcal{R}(V)$ it holds that $w \in \mathcal{R}(V')$, where $V' = \{v_1, \ldots, v_{n-1}, v'\}$ and $v' = w$.

**Table 1.** Summary of our main results. Each block of rows corresponds to a setting, with rows in the block corresponding to aggregation methods.

| Setting | Aggregation | Solution | Unique | Complexity | Majoritarian | Monotonicity |
|---|---|---|---|---|---|---|
| Plurality | Condorcet | Plurality | no | linear | yes | yes |
| Elections | $\widetilde{L}_p(V)$ | Plurality | no | linear | yes | yes |
| 1D | Condorcet | Median | for $n \in \mathbb{N}_{odd}$ | linear | yes | yes |
| Single | $\widetilde{L}_p(V)$ | Median for $p=1$ | yes | linear for $p=1,2,\infty$ | only for $p=1$ | only for $p=1$ |
| Winner | | Average for $p=2$ | | efficient for $p \neq 1,2,\infty$ | | |
| | | Mid-range for $p=\infty$ | | | | |
| Continuous | Condorcet | no | no | ? | yes | yes |
| Budgeting | $\widetilde{L}_p(V)$ | yes | yes | linear for $p=2$ | only for $p=1$ | no |
| | | | | efficient for $p \neq 2$ | | |
| Social | Condorcet | no | no | ? | yes | yes |
| Welfare | $\widetilde{L}_p(V)$ | yes | no | NP-hard for $p=1,\infty$ | only for $p=1$ | only for $p=1$ |
| Functions | | | | | | |
| Committee | Condorcet | no | no | ? | yes | yes |
| Elections | $\widetilde{L}_p(V)$ | yes | for $p=1, n \in \mathbb{N}_{odd}$ | linear for $p=1$ | only for $p=1$ | only for $p=1$ |
| | | | | NP-hard for $p=\infty$ | | |
| Participatory | Condorcet | no | no | ? | yes | yes |
| Legislation | $\widetilde{L}_p(V)$ | yes | no | NP-hard for $p=1,\infty$ | only for $p=1$ | only for $p=1$ |

**Settings.** For each setting, we specify a corresponding metric space $(X, d)$: (1) Plurality elections with alternatives $A$ corresponds to $X = A$ and $d(x, y) = 1$ for $x \neq y$ and $d(x, x) = 0$. (2) 1-dimensional elections with alternatives $X \subset \mathbb{R}$ corresponds to $X = A$ and $d(x, y) = |x - y|$. (3) Continuous participatory budgeting, where a dollar is to be split among $m$ alternatives, corresponds to $X = \{(v_1, \ldots, v_m) \in \mathbb{R}^m \mid v_i \geq 0 \text{ and } \sum_i v_i = 1\}$ and $d(x, y) = \|x - y\| = \sqrt{\sum_i |x_i - y_i|^2}$. (4) Social welfare functions corresponds to $X$ being the set of relevant permutations with swap distance $d$. (5) Committee elections, where a subset of $A$ is to be elected, corresponds to $X \subseteq 2^A$ and $d(x, y)$ is the symmetric difference between $x$ and $y$. (6) Participatory legislation, where strings of size $\leq \ell$ over an alphabet $\Sigma$ are to be aggregated into a single string, corresponds to $X = \Sigma^*$ and $d(x, y)$ is a weighted-Levenshtein distance, i.e., the minimum cost of transforming $x$ to $y$, where *insert/delete* cost 1 and a *swap* costs $1/\ell^2$.

# References

1. Anshelevich, E., Bhardwaj, O., Elkind, E., Postl, J., Skowron, P.: Approximating optimal social choice under metric preferences. AI **264**, 27–51 (2018)
2. Elkind, E., Faliszewski, P., Slinko, A.: Distance rationalization of voting rules. Soc. Choice Welfare **45**(2), 345–377 (2015)
3. Enelow, J.M., Hinich, M.J.: The spatial theory of voting: An introduction (1984)
4. Feldman, M., Fiat, A., Golomb, I.: On voting and facility location. In: Proceedings of the 2016 ACM Conference on Economics and Computation, pp. 269–286 (2016)
5. Meir, R., Procaccia, A.D., Rosenschein, J.S.: Algorithms for strategyproof classification. Artif. Intell. **186**, 123–156 (2012)
6. Procaccia, A.D., Rosenschein, J.S.: The distortion of cardinal preferences in voting. In: Proceedings of CIA '06, pp. 317–331 (2006)
7. Shahaf, G., Shapiro, E., Talmon, N.: Aggregation over metric spaces: Proposing and voting in elections, budgeting, and legislation. arXiv preprint arXiv:1806.06277 (2018)
8. Shapiro, E.: Point: foundations of e-democracy. Commun. ACM **61**(8), 31–34 (2018)
9. Skowron, P., Elkind, E.: Social choice under metric preferences: scoring rules and STV. In: Proceedings of AAAI '17, pp. 706–712 (2017)

# Application of Boolean Logic to Natural Language Complexity in Political Discourse

Austin Taing$^{(\boxtimes)}$, Judy Goldsmith, and Justin Wedeking

University of Kentucky, Lexington, KY 40506, USA
austin.taing@uky.edu

This study employs several well-studied measures of linguistic complexity and proposes a new one to examine whether politicians change their language to become more or less difficult to parse in different situations. We use 27,500 press releases from the US Senate between 2004–2008 and examine election cycles and natural disasters (hurricanes) as situations in which politicians' language may change. We calculate the syntactic complexity and readability of each press release, and propose a new measure, *logical complexity*, to investigate classical Boolean logic as a measure of linguistic complexity. In our sample, language becomes more complex in coastal senators' press releases concerning hurricanes, but we see no significant change for election cycles. Our measure shows similar results to those of the well-established ones, showing that logical complexity is a useful lens for measuring linguistic complexity.

**Experimental Setup.** We employ two categories of well-studied linguistic complexity measures: syntactic complexity and readability. They establish a baseline for comparison with our logical complexity measure. The syntactic complexity measures used are *clauses per sentence*, *T-unit length*, and *complex T ratio*. We use T-units as defined by Hunt [3] as a base information unit for the text. These measures focus on the amount of information presented per unit. The readability measures used are *Flesh Reading Ease* [1] and *Automated Readability Index* [4], which represent more theoretical and practical approaches to text readability, respectively.

We propose and test *logical complexity*, defined as the number of logical operation keywords per T-unit in a given text. This treats each T-unit as a logical predicate and seeks to identify how many elements are present in each; this represents the length of a Boolean formula representing that T-unit. This study uses a collection of transcripts of press releases from the offices of the US Senate from 2004–2008 [2]. We examine two groups of senators: the election group and the coastal group. The election group consists of senators up for re-election in 2006, and compares their press releases during the election cycle of that year, January through October, with the rest of their press releases. The coastal group consists of senators of states along the Southeast coast which saw significant hurricanes during the years represented, comparing press releases concerning hurricanes (identified by keywords) and those not.

**Results.** Using the syntactic complexity and readability measures, we found no significant trend in language complexity changes for the election group. The

© Springer Nature Switzerland AG 2019
S. Pekeč and K. B. Venable (Eds.): ADT 2019, LNAI 11834, pp. 175–176, 2019.
https://doi.org/10.1007/978-3-030-31489-7

coastal group showed a consistent trend of increased complexity in hurricane-related press releases across the readability measures and T-unit length. These graphs are shown in Fig. 1. However, this trend did not extend to the other syntactic complexity measures. The results of our logical complexity measure were very similar to those of the other measures.

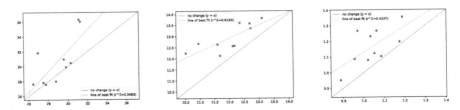

**Fig. 1.** T-length, ARI, and logical complexity for coastal group

**Conclusion.** Our experiment tested the logical complexity against well-known measures of linguistic complexity. We used all of the measures to investigate whether politicians' language becomes simpler when discussing natural disasters or during election cycles. f We saw that our logical complexity measure gave similar results to well-established complexity measures, especially the readability measures, which is fairly surprising; our definition of logical complexity uses structures of syntactic complexity, T-units, rather than the more general structures employed by readability measures. These results imply that our measure does reasonably well in determining the complexity of text. We assert then that while this study serves primarily as a proof of concept, the results suggest that Boolean logic can serve as a useful basis for practical examination of language complexity.

# References

1. Flesch, R.: A new readability yardstick. J. Appl. Psychol. **32**(3), 221 (1948)
2. Grimmer, J.: Replication data for: Representational Style: The Central Role of Communication in Representation (2010). 1902.1/14596, https://hdl.handle.net/1902.1/14596
3. Hunt, K.W.: Grammatical Structures Written at Three Grade Levels. NCTE Research Report No. 3. (1965)
4. Senter, R., Smith, E.A.: Automated Readability Index. Cincinnati University, OH, Tech. rep. (1967)

# Integrating Multiple Contexts into Multi-criteria Majority-Rule Sorting

Arthur Valko[1,2(✉)], Alexandru-Liviu Olteanu[3], and Patrick Meyer[2]

[1] Chaire de cyberdéfense des systémes navals, École navale, Lanvéoc, France
arthur.valko@ecole-navale.fr
[2] IMT Atlantique, Lab-STICC UMR CNRS 6285, 29238 Brest, France
[3] Université Bretagne Sud, Lab-STICC, UMR 6285, CNRS, Lorient, France

**Keywords:** Multi-criteria decision aiding · Context-dependent evaluations and preferences · Majority-rule sorting · Mixed-integer programming

The field of Multi-Criteria Decision Aiding (MCDA) seeks to help decision-makers (DMs) facing difficult decisions when multiple, often conflicting, criteria are considered [3]. Evaluations of alternatives on multiple criteria may vary according to different application contexts. However, in MCDA, decision alternatives are usually evaluated on multiple criteria for a specific context [1]. The preferences of the DM may also be considerably different for each of these contexts. To provide more accurate recommendations, the decision aiding process take these contexts into account.

We focus on the sorting problem in MCDA, and propose an extension of MR-Sort [2,4] to deal with alternatives evaluated on multiple criteria and across multiple contexts, which we call MR-Sort-C. The proposed approach considers each context as an MR-Sort sub-problem, providing an intermediate classification of an alternative with respect to it, and then uses these classifications as evaluations for an overall MR-Sort model, which then provides the final classification.

Let us consider a finite set of alternatives $A$, a finite set of criteria indexes $J = \{1, ..., m\}$ and a finite set of context indexes $T = \{1, ..., n\}$.

For each MR-Sort sub-model (one for each $t \in T$), we define a majority threshold $\lambda_t \in ]0.5, 1]$, a set of criteria weights $\underline{w}_{t,j} \in [0,1], \forall j \in J$ with $\sum_{j \in J} \underline{w}_{t,j} = 1$, and category profiles evaluations $\underline{b}_{t,h,j}, \forall t \in T, \forall h \in 1..k_t + 1, \forall j \in J$. Each category $\underline{c}_{t,h}, t \in T, h \in K_t$ of the $t^{\text{th}}$ sub-model is delimited in the criteria space by a lower frontier $\underline{b}_{t,h-1}$ and an upper one $\underline{b}_{t,h}$. Furthermore the frontier performances are non-decreasing. i.e . $\underline{b}_{t,h,j} \geqslant \underline{b}_{t,h-1,j} \forall t, h, j \in T \times \{1..k_t + 1\} \times J$. Two rules to assign an alternative to a class may be found in the literature: the pessimistic and the optimistic rules. We will use the first as it is the most commonly used. An alternative $a$ is therefore assigned to the highest possible category such that $a$ outranks its lower frontier but not its upper frontier. An alternative $a$ is said to outrank another (in our case a category limit) if and only

© Springer Nature Switzerland AG 2019
S. Pekeč and K. B. Venable (Eds.): ADT 2019, LNAI 11834, pp. 177–179, 2019.
https://doi.org/10.1007/978-3-030-31489-7

if there is a sufficient coalition of criteria supporting the assertion that $a$ is at least as good as the other alternative. For each sub-model with $t \in T$, we define local concordance indices $(\underline{C}_{t,j}(a, b_{t,h}))$ between an alternative $a$ and a category profile $b_{t,h}$, as well as a global one $(\underline{C}_t(a, b_{t,h}))$:

$$\underline{C}_{t,j}(a, b_{t,h}) = \begin{cases} 1, \text{ if } g_{t,j}(a) \geqslant \underline{b}_{t,h,j} \\ 0, \text{ otherwise.} \end{cases} \qquad \underline{C}_t(a, b_{t,h}) = \sum_{j \in J} \underline{w}_{t,j} \underline{C}_{t,j}(a, b_{t,h}) \quad (1)$$

In order to determine whether they are sufficient in order to validate an outranking situation, we compare global concordances to the majority thresholds $\underline{\lambda}_t$. We define for each sub-model the $\underline{S}_t$ relations as:

$$a \, \underline{S}_t \, \underline{b}_{t,h} \iff \underline{C}_t(a, \underline{b}_{t,h}) \geqslant \underline{\lambda}_t, \forall t \in T, \forall h \in 1..k_t + 1 \qquad (2)$$

Using these outranking relations, we can now define the assignment rule of the MR-Sort sub-models. An alternative $a$ is assigned to category $\underline{c}_{t,h}, \forall t \in T$. The results of these assignments, for all sub-models $t \in T$, will then form the aggregated evaluations of the alternatives, which will then be used in the global model:

$$a \in \underline{c}_{t,h} \iff a \, \underline{S}_t \, \underline{b}_{t,h} \text{ and } a \, \underline{\mathcal{S}}_t \, \underline{b}_{t,h+1} \qquad g_t(a) = h \iff a \in \underline{c}_{t,h}, \forall t \in T \quad (3)$$

We define the overall MR-Sort model using a majority threshold $\bar{\lambda} \in ]0.5, 1]$, criteria weights $\bar{w}_t \in 0, 1, \forall t \in T$ with $\sum_{t \in T} \bar{w}_t = 1$, and category profiles evaluations $\bar{b}_{t,h}, \forall t \in T, \forall h \in 1..k + 1$. Similarly to the MR-Sort sub-models, each category $\bar{c}_h, h \in \{1, ..., k\}$ of the top-model is delimited in the criteria space by a lower frontier $\bar{b}_{h-1}$ and an upper one $\bar{b}_h$ and the frontier performances are non-decreasing. i.e . $\bar{b}_{h,t} \geqslant \bar{b}_{h-1,t}, \forall h, t \in \{1..k_t + 1\} \times T$. We define a local concordance index $\bar{C}_t$, a global concordance index $\bar{C}$ and an outranking relation $\bar{S}$, replacing the model parameters with the ones defined above and the alternative evaluations with the ones in Equation 3:

$$\bar{C}_t(a, \bar{b}_h) = \begin{cases} 1, \text{ if } g_t(a) \geqslant \bar{b}_{h,t} \\ 0, \text{ otherwise.} \end{cases} \qquad \bar{C}(a, \bar{b}_h) = \sum_{t \in T} \bar{w}_t \bar{C}_t(a, \bar{b}_h) \qquad (4)$$

$$a \, \bar{S} \, \bar{b}_h \iff \bar{C}(a, \bar{b}_h) \geqslant \bar{\lambda}, \forall h \in 1..k + 1 \qquad (5)$$

An alternative will then be assigned to a category $\bar{c}_h$ if and only if $a \, \bar{S} \, \bar{b}_h$ and $a \, \bar{\mathcal{S}} \, \bar{b}_{h+1}$. This step gives the overall assignment of the alternative, according to the MR-Sort-C model.

An indirect inference approach for constructing the entire MR-Sort-C model from assignment examples based only on the overall category has also been developed using a mixed-integer linear program. The algorithm has been tested over artificially constructed benchmarks, however, due to its complexity, may only be used on small problem instances. Nevertheless, the precision of the model is at least as good as that of the classical MR-Sort model (for which the criteria have been duplicated along the various contexts). Compared to that model, the MR-Sort-C model also provides more readable descriptions of the final assignments, through a hierarchical perspective.

# References

1. Bisdorff, R., Dias, L.C., Meyer, P., Mousseau, V., Pirlot, M. (eds.): Evaluation and Decision Models with Multiple Criteria. IHIS. Springer, Heidelberg (2015). https://doi.org/10.1007/978-3-662-46816-6
2. Leroy, A., Mousseau, V., Pirlot, M.: Learning the Parameters of a Multiple Criteria Sorting Method. In: Brafman, R.I., Roberts, F.S., Tsoukiàs, A. (eds.) ADT 2011. LNCS (LNAI), vol. 6992, pp. 219–233. Springer, Heidelberg (2011). https://doi.org/10.1007/978-3-642-24873-3_17
3. Roy, B.: Multicriteria Methodology for Decision Aiding, pp. 1257–8. Kluwer Academic, Dordrecht (1997)
4. Sobrie, O., Mousseau, V., Pirlot, M.: Learning a Majority Rule Model from Large Sets of Assignment Examples. In: Perny, P., Pirlot, M., Tsoukiàs, A. (eds.) ADT 2013. LNCS (LNAI), vol. 8176, pp. 336–350. Springer, Heidelberg (2013). https://doi.org/10.1007/978-3-642-41575-3_26

# Author Index

Printed in the United States
By Bookmasters